Test Equipment

Test Equipment

AVO Multi-Amp Institute

AVO Multi-Amp® Institute

Delmar Publishers Inc.™

I T P

Notice to the reader

Publisher does not warrant or guarantee any of the products described herein or perform any independent analysis in connection with any of the product information contained herein. Publisher does not assume, and expressly disclaims, any obligation to obtain and include information other than that provided to it by the manufacturer.

The reader is expressly warned to consider and adopt all safety precautions that might be indicated by the activities described herein and to avoid all potential hazards. By following the instructions contained herein, the reader willingly assumes all risks in connection with such instructions.

The publisher makes no representations or warranties of any kind, including but not limited to, the warranties of fitness for particular purpose or merchantability, nor are any such representations implied with respect to the material set forth herein, and the publisher takes no responsibility with respect to such material. The publisher shall not be liable for any special, consequential or exemplary damages resulting, in whole or in part, from the readers' use of, or reliance upon, this material.

Cover design by Cheri Plasse
Front cover reproduced with permission from John Fluke Mfg. Co., Inc.
Back cover photo courtesy of AVO Megger®
Freelance Project Editor: Pamela Fuller

Delmar Staff
Senior Executive Editor: Mark Huth
Developmental Editor: Sandy Clark Gnirrep
Senior Project Editor: Laura Gulotty
Project Editor: Elena Mauceri
Production Coordinator: Dianne Jensis
Art/Design Coordinators: Brian Yacur/Cheri Plasse

For information, address Delmar Publishers Inc.
3 Columbia Circle, Box 15-015
Albany, New York 12212-5015

Printed in the United States of America
Published simultaneously in Canada by Nelson Canada,
a division of The Thomson Corporation

1 2 3 4 5 6 7 8 9 10 xxx 00 99 98 97 96 95 94

Library of Congress Cataloging-in-Publication Data

Test equipment / Multi-Amp Institute.
    p. cm.
  Includes index.
  ISBN 0-8273-4923-8
  1. Electronic instruments. 2. Measuring instruments. I. Multi-Amp Institute.
TK7878.4.T45 1994
  621.37—dc20   93-20199
                                          CIP

Contents

Chapter 2

Chapter 4
Care and Application of Meter Test Equipment

Chapter 7
Watt, Watthour Phase-Angle Meters   155

Chapter 8
Motor, Cable, and Transformer Test Equipment   183

Chapter 9
Circuit Breaker Test Sets   213

Preface

Test Equipment is designed to expose the reader—whether student, electrical utility worker, or industrial electrician—to a variety of test equipment found today in the electrical industry.

Test Equipment takes nothing for granted; it begins with a review of the electrical and mathematics skills needed for the use and application of test equipment. Continuing with analog and digital measuring devices, *Test Equipment* progresses through a range of laboratory and field test equipment to heavy duty test sets for testing industrial circuit breakers.

Each chapter begins with a list of objectives. The objectives serve as a preview of what the reader will learn from the chapter, as well as a gauge of what is expected of him. Test questions at the end of each chapter measure the user's comprehension of the material presented, and reinforce the more important concepts of each chapter.

Test Equipment serves either as a teaching tool for the student or as a review of operating principles and applications for the electrical technician. Its purpose is to compliment rather than replace the manufacturer's equipment instruction manual.

This book works equally well as a self-teaching guide, or for use by an instructor as a text for an industrial instrumentation class. Chapters progress from simple measuring elements through use and applications of these devices in measuring instruments. Instruments covered increase in complexity, so that the instructor can gauge by student reaction where emphasis is necessary.

An additional benefit of *Test Equipment* is its use in decisions about test equipment acquisitions. It gives the information needed to decide whether the equipment described will perform the job required.

Acknowledgements

AVO Multi-Amp Institute and Bob Depta wish to thank these companies and individuals for their invaluable assistance in the preparation of this book:

Sandy Young	AVO Multi-Amp Institute
Robert Riordan	Allen Bradley Company
Kenneth L. Carl	A.B. Chance Company
Kevin Basso	A.W. Sperry Instruments Inc.
Robert Christopher	Amprobe Instrument®
Peg O'Neal	AVO Biddle Instruments
Valerie Sawers	AVO Megger
George Gore	B+K Precision
Eric Forman	EMCO Division, Components Specialities, Inc.
Ron Brittain	John Fluke Manufacturing Company, Inc.
Randy S. Rowan	H D Electric Company
Michelle Phillips	IEEE
Robert M. Dunbar	Knopp Inc.
Lynn Feiner	North Hand Protection
Harry C. Ploehn	Ploehn Engineering and Consulting
Joseph E. Deans	Shalltronix Corporation
Ron Barma	Simpson Electric Company
Lowell Bonnett	Square D Company
Jack Schramm	Staco Energy Products Company
Susan Hancock	TIF Instruments, Inc.
Warren Hess	Triplett Corporation

Foreword

AVO Multi-Amp Institute, a division of AVO Multi-Amp Corporation, has provided training to students from industry, utilities, government, and military facilities worldwide since 1963. At the learning center in Dallas, and regional learning centers in Orlando, Philadelphia, Toronto, and sites around the world, AVO Multi-Amp Institute has trained over 70,000 students.

This textbook was written using the same Instructional Systems Development (ISD) method used to produce more than 80 technician-level maintenance training programs at the Institute. AVO Multi-Amp technical writers and instructors are field-experienced, and they target instruction for job performance requirements.

AVO Multi-Amp Corporation is the pre-eminent provider of high quality test equipment, testing and maintenance services, and technical training for electrical power equipment and power systems. Its products, especially in the area of high current test sets, join with those of other leading test equipment manufacturers to provide a comprehensive overview of test instruments found in most electrical maintenance shops.

In developing this textbook, Robert P. Depta draws upon his forty years of experience in the field of electrical instrumentation. As Senior Training Specialist with AVO Multi-Amp Institute, he teaches a wide range of power system training courses, including Basic Electricity, Electrical Print Reading, Basic and Advanced Watthour Meter Maintenance, Protective Relay Maintenance, and Solid-State Watthour Meter Maintenance.

Depta holds a Bachelor of Science degree in education from Seton Hall University. He founded Metering Services Company, a consulting firm that installed and repaired electrical instrumentation equipment in the New Jersey area.

For eleven years Mr. Depta served as an application engineer and product manager for Multi-Amp meter test equipment. His experience also includes 27 years of extensive lab, shop, and field expertise in the application, testing, and maintenance of electrical test equipment.

List of Figures and Tables

Chapter 4

Chapter 8

Review of Electrical Theory   1

INTRODUCTION

Understanding a few basics of electrical theory makes it much easier to learn the safe and proper use of electrical test equipment. This chapter offers a brief review of those theories. It also offers examples for solving the simple electrical circuits used in most measuring instruments. Before going on to Chapter 2, invest a few moments of your time by reviewing this material. It will be well worth your while.

OBJECTIVES

After studying this chapter, the student should be able to:

- *Understand how electrons cause the flow of electricity.*
- *Describe DC voltage.*
- *Explain the relationship between voltage, current, and resistance by using Ohm's Law.*
- *Explain what power is.*
- *Understand series and parallel circuits.*
- *Explain how AC voltage is generated.*
- *Know the difference between peak, rms, and average voltage.*
- *Explain the OSHA ruling on working with voltage.*

DC CURRENT

Two forms of voltage are in general use today: DC voltage, a steady-state voltage, and AC, which periodically changes direction. Direct current is what you expect to get out of a battery, such as those used to provide current and voltage to operate automobiles. If you look at a chart showing the voltage output of the battery over a period of time—one minute, for example—it will show a straight line at approximately 12 volts, Figure 1–1. With DC, the voltage level does not change.

In the paragraph above DC is defined as direct current, which is a measure of amperes. Further, we noted that an automobile battery will produce 12 volts. To better understand what is happening, perhaps it is best to review the conditions that go on in an electrical circuit.

A conductor, such as a piece of copper wire, is made up of millions of copper molecules. The molecules in turn are made of atoms. Each atom contains protons, a positive charge located in the center. In rings spinning around the proton are electrons, a negative charge, as shown in Figure 1–2. Good conductors of electricity, such as aluminum or copper, have electrons in their outer rings, called free electrons, that are capable of moving from one atom to another.

If a force is applied across a complete circuit, from a battery or generator, free electrons will move from one atom to the next. When electrons move, electricity is created.

Two theories about the direction the electrons will flow are the *conventional* and *electron* theories. Conventional theory, which originated back in the days of Ben Franklin in the 1700s, holds that electrons flow from the positive to the negative. The more modern electron theory says that electrons flow from the negative to the positive. In this textbook, the electron theory, with flow from negative (-) to positive (+), will be used.

If a wire is connected across the plus and minus terminals of the battery, as shown in Figure 1–3, electrons will certainly move. In fact, in this circuit, called a short circuit, the electrons within the wire will have unrestricted flow. Soon either the wire will overheat or the battery will be exhausted and no longer able to move electrons. Some means must be used to control or resist this electron flow.

FIGURE 1–1
Direct current

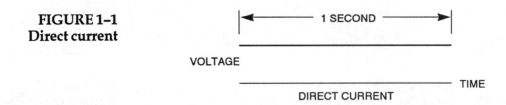

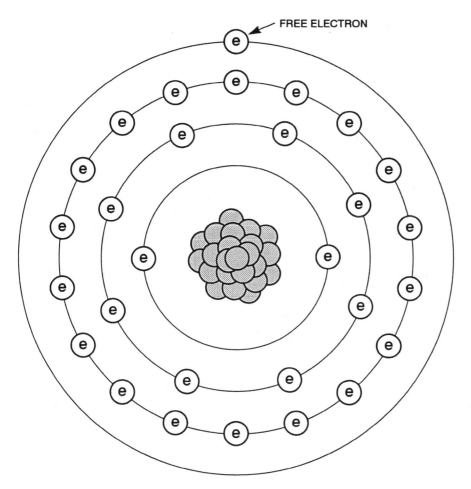

FREE ELECTRON

FIGURE 1–2
Copper atom

If the same copper conductor is used to wire a bulb across the terminals of the battery, as shown in Figure 1–4, the bulb will light. Made of high resistance wire, the bulb filament heats up as it opposes and controls the electrons flowing through it. Not only is light given off, but battery life will also be much longer.

RELATIONSHIP BETWEEN VOLTAGE, CURRENT, AND RESISTANCE

Three measurable elements exist in the simple battery, conductor, and bulb circuit. They are voltage, current, and resistance.

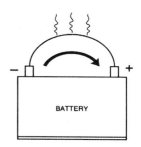

BATTERY

FIGURE 1–3
Battery with wire

Voltage

Voltage is electrical pressure, and is often compared to pressure supplied by a pump in a water system. Most important for the study of test instruments is electrical pressure produced by chemicals (battery), magnetism (generator), or heat (thermocouple). Electrical formulas use the letter (E) for voltage.

Current

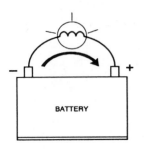

FIGURE 1–4
Battery with bulb

Measured in amperes, current is the number of electrons that flow past a given point in one second. How many electrons? Write 6,250 and then add 15 more zeroes. The letter (I) is used to identify current.

Resistance

Measured in ohms, resistance is a substance or device that opposes the flow of current. Some examples of resistors are the special nichrome (nickel-chrome) wire used in electric toasters to generate heat, or carbon composition resistors like those found in many electronic circuits. The letter "R" identifies resistance in electrical formulas, while the symbol "Ω" is used in most schematics.

It is obvious from the circuit shown in Figure 1–4 that there is a direct relationship between voltage, amperes, and ohms. Perhaps the definition for the volt sums it up best: A volt is the pressure necessary to push one ampere through one ohm of resistance.

Ohm's Law

This definition leads into Ohm's Law, which states:

$$\frac{E}{IR}$$

By covering up the quantity you wish to know, the formula will show whether you need to multiply or divide the remaining quantities to solve the problem.

Some examples of how Ohm's Law works.

$$E = 12$$
$$I = 2$$
$$R = 6$$

Solve for volts.
　　Covering up the E, for volts, leaves:

　　　I R

Given I = 2 Amperes, and R = 6 Ohms: E = I × R = 2 × 6 = 12 Volts

Solve for amps.
　　Covering up the I, for amps, leaves:

$$\frac{E}{R}$$

Given E = 12 and R = 6 Ohms: $I = \frac{E}{I} = \frac{12}{6} = 2$ Amperes

Solve for ohms.
　　Covering up the R, for ohms, leaves:

$$\frac{E}{I}$$

Given E = 12 and I = 2: $R = \frac{E}{I} = \frac{12}{2} = 6$ Ohms

Power

One more electrical term must be mentioned, and that is *power*. Power is defined as a measurement of the rate work is being done, measured in watts. When one volt pushes one ampere of current through one ohm of resistance, some heat is given off, and the result is one watt of power. The work, in this case, is the heat created.
　　The formula for power happens to be the same for both DC or AC voltage, in a purely resistive load, and is equal to:

　　E × I

Example:

Volts = 10
Amperes = 2
Power = E × I = 10 × 2 = 20 Watts

Suppose you want to solve for watts using the above formula, but you do not have a voltage reading. But you know the circuit has a total resistance of 5 ohms. The formula for volts is E × I, and you can, of course, solve for volts, and then plug that value into the power formula. A simpler way is to combine the formula for volts with the formula for power in the following manner:

Begin with the formula for volts (I × R). Now that volts are determined, finish the power formula by adding current (I), and you end up with:

I × R × I

Since there are two current values in the formula, you must multiply the current times the current. Another way of showing this in the formula is by writing the current symbol (I) and adding an exponent, I^2. The exponent states the number of times you must multiply the number by itself. The formula now becomes:

Watts = I^2 × R or I^2R, and equals 2 × 2 × 5 = 20 Watts

Now, supposing you wanted to solve the above problem but did not have the current reading. From Ohm's Law, current equals :

$$\frac{E}{R}$$

Once again combine the formula for amps with the formula for power:

$$Watts = \frac{E}{R} \times E$$

There are two voltage values in the formula, so voltage must be multiplied times voltage. To simplify, again use an exponent and the formula is:

$$Watts = \frac{E^2}{R} = \frac{10 \times 10}{5} = \frac{100}{5} = 20 \ Watts$$

SERIES CIRCUITS

A DC circuit that has a single path of current flow is called a series circuit, as shown in Figure 1–5. Three laws govern how current, voltage, and resistance react in a series circuit. If the load is totally resistive,

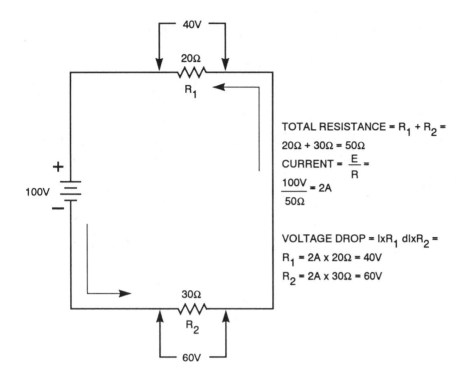

FIGURE 1-5
Series circuit

TOTAL RESISTANCE = $R_1 + R_2$ =

$20\Omega + 30\Omega = 50\Omega$

CURRENT = $\dfrac{E}{R}$ =

$\dfrac{100V}{50\Omega}$ = 2A

VOLTAGE DROP = IxR$_1$ dIxR$_2$ =

R_1 = 2A x 20Ω = 40V

R_2 = 2A x 30Ω = 60V

as will be the case with most analog measuring instruments, these laws apply to both AC and DC circuits.

1. Current will measure the same at any point in the circuit.
2. Voltage drops across the individual resistors will add up to the total applied voltage.
3. The additive total of individual resistors will equal the total circuit resistance.

Example:

R_1= 20 Ohms
R_2= 30 Ohms
Voltage = 100 Volts
Total resistance = $R_1 + R_2$ = 20 + 30 = 50 Ohms

Using Ohm's Law, current in the circuit equals $\dfrac{100V}{50}$ = 2 A

Voltage drop across resistor R_1 = 20 x 2A = 40 Volts
Voltage drop across resistor R_2 = 30 × 2A = 60 Volts

Total voltage dropped across all resistors is 40 plus 60, or 100 volts, which matches input voltage.

PARALLEL CIRCUITS

If a circuit has more than one current path, it is called a parallel circuit. Figure 1–6 shows such a circuit, which is governed by three basic laws concerning voltage, current, and resistance. Once again, in a purely resistive circuit, these laws apply to both AC and DC.

1. The voltage in a parallel circuit will be the same throughout each branch of that circuit.
2. Current flow through each branch of a circuit will be dependent on the resistance of that branch, and total current flow will be equal to the sum of currents flowing in all branches.

FIGURE 1–6
Parallel circuit

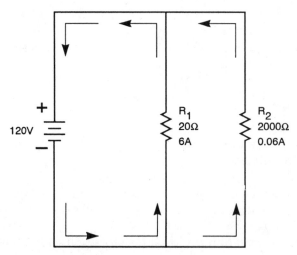

Total resistance reciprocal method

$$\frac{1}{\dfrac{1}{R_1} + \dfrac{1}{R_2}} = \frac{1}{\dfrac{1}{20} + \dfrac{1}{2000}} = \frac{1}{0.05 + 0.0005} = \frac{1}{0.0505} = 19.8\ \Omega$$

Total current =

$$I = \frac{E}{R} = \frac{120\ V}{19.8\ \Omega} = 6.06\ A$$

Current through each resistor =

$$I\ in\ R_1 = \frac{120\ V}{20\Omega} = 6\ A$$

$$I\ in\ R_2 = \frac{120\ V}{2000\ \Omega} = 0.06\ A$$

Total resistance product/sum =

$$\frac{R_1 \times R_2}{R_1 + R_2} = \frac{20 \times 2,000}{20 + 2,000} = \frac{40,000}{2,020} = 19.8\ \Omega$$

3. Total resistance in a parallel branch is always less than or equal to the smallest resistor making up that branch.

Example:

R_1 = 20 Ohms
R_2 = 2,000 Ohms
Volts = 120

In a parallel circuit, two methods are most commonly used to solve for total circuit resistance.

Reciprocal Method

Using reciprocals is just another mathematical tool for solving problems. The reciprocal of a number is that number divided into one (1). For example, the reciprocal of 25 is 1/25 and equals 0.04.

In the reciprocal method, the reciprocals of all resistors are found, and in turn made into a reciprocal:

$$\cfrac{1}{\cfrac{1}{R_1} + \cfrac{1}{R_2}} = \cfrac{1}{\cfrac{1}{20} + \cfrac{1}{2000}} = \frac{1}{0.5 + .0005} = \frac{1}{.0505} = 19.8 \text{ Ohms}$$

The reciprocal method will work for any number of resistors in parallel.

Product Over the Sum Method

This method is a less confusing way to solve problems when only two resistors are involved; but with proper manipulation, total resistance for any number of resistors in a parallel circuit can be found.

$$\text{Total resistance} = \frac{R_1 \times R_2}{R_1 + R_2} = \frac{20 \times 2000}{20 + 2000} = \frac{40000}{2020} = 19.8 \text{ Ohms}$$

Once total resistance in the circuit is known, solve for current.

$$I = \frac{E}{R} = \frac{120V}{19.8 \text{ Ohms}} = 6.06 \text{ A}$$

When working with electrical circuits, and using voltmeters, ammeters, and ohmmeters, it is often necessary to use smaller or larger units than the volt, ampere, and ohm. Very often millivolts, microamperes, or

Amperes

One ampere = 1000 Milliamperes (mA)= 1000000 Microamperes (µA)

Calculator would read

.000001 = 1 Microampere	1×10^{-6}
.000010 = 10 Microamperes	10×10^{-6}
.000100 = 100 Microamperes	100×10^{-6}
.001000 = 1000 Microamperes = 1 Milliampere	1×10^{-3}
.010000 = 10 Milliamperes	10×10^{-3}
.100000 = 100 Milliamperes	100×10^{-3}
1.000000 = 1000 Milliamperes = 1 Ampere	

Volt

One volt = 1000 Millivolts (mV) = 1000000 Microvolts (µV)

Calculator would read

.000001 = 1 Microvolt	1×10^{-6}
.000010 = 10 Microvolts	10×10^{-6}
.000100 = 100 Microvolts	100×10^{-6}
.001000 = 1000 Microvolts = 1 millivolt	1×10^{-3}
.010000 = 10 Millivolts	10×10^{-3}
.100000 = 100 Millivolts	100×10^{-3}
1.000000 = 1000 Millivolts = 1 Volt	
1000 = 1k Volt	1×10^{3}

Ohms

Calculator would read

1000 = 1k Ohms	1×10^{3}
1000000 = 1M (meg) Ohms	1×10^{6}

Watts

Calculator would read

1000 = 1 Kilowatt	1×10^{3}

megohms come into play. The chart on the previous page shows the relationship between these values and how they often appear on an electronic calculator in what is called scientific notation.

To change a number in scientific notation to its decimal equivalent, use the following simple procedures:

Notice that in scientific notation the numbers are expressed as powers of 10 with some exponent (the number to the right and slightly higher than the 10). For example, 100 microamperes is given as 100×10^{-6}. The negative exponent means the answer will be less than one. To change to a decimal, first write down the 100 and put the decimal point where it should be for a whole number.

100.

Now count off to the left the number of places as shown by the negative exponent, adding zeros as place holders.

.000100.

The new decimal place is now on the left, and the number becomes .0001.

If the exponent is a positive number, the procedure is almost the same, except that the decimal is moved to the right. From the chart, one million ohms = 1×10^{6}, write the one and include the decimal.

1.

Count off the decimal places as called for by the exponent and add zeroes as place holders to the right.

1.000000.

Remove the original decimal point, and you have 1,000,000, or 1 million ohms (1 meg).

CURRENT FLOW AND MAGNETISM

Two theories of electron current flow are in use today. Conventional current flow states that electrons flow from the positive side of a circuit to the negative. The second theory, electron current flow, states that electrons flow from the negative to the positive side of the circuit. In this textbook, only electron current flow will be used.

When current flows through a conductor, it sets up a magnetic field around it. This field is called magnetic flux. The direction in which flux lines flow is directly related to the direction in which current flows. A simple way of finding flux direction is called the *left hand magnetic field*

FIGURE 1–7
Left-hand magnetic
field rule

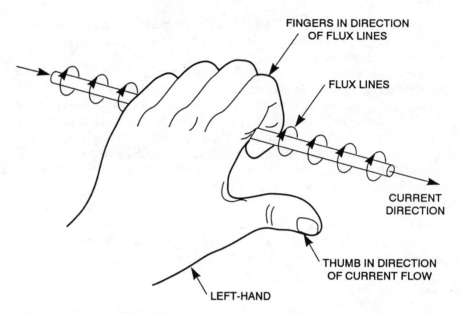

FINGERS IN DIRECTION
OF FLUX LINES

FLUX LINES

CURRENT
DIRECTION

THUMB IN DIRECTION
OF CURRENT FLOW

LEFT-HAND

rule, Figure 1–7. If the conductor is grasped in the left hand with the thumb pointing in the direction of current flow—minus to plus—the fingers will point in the direction of flux lines.

When a current-carrying conductor is positioned between the poles of a permanent magnet, the flux fields from both will interact, as shown by the flux arrows in Figure 1–8. On the side of the conductor where both flux fields point in the same direction, they strengthen each other. On the opposite side of the conductor, the magnetic fields oppose, and their flux strength is weakened. As a result the conductor is forced from the more powerful flux field in the direction where flux strength is the weakest. Called motor action, this force is the basis for meter move-

FIGURE 1–8
Current-carrying
conductor in
magnetic field

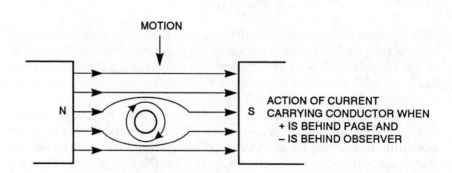

MOTION

N   S

ACTION OF CURRENT
CARRYING CONDUCTOR WHEN
+ IS BEHIND PAGE AND
− IS BEHIND OBSERVER

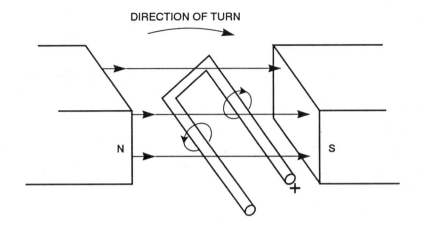

DIRECTION OF TURN

FIGURE 1–9
Current-carrying coil
in magnetic field

ments and becomes more intense should the conductor be wound in the shape of a coil, Figure 1–9.

Motor action is just another way of saying unlike poles attract; like poles repel.

ALTERNATING CURRENT

In the study of DC voltage, it was shown that a conductor suspended in a magnetic field will be pushed aside if a current is passed through it. It is also true that, if a conductor is moved through a magnetic field or if a magnetic field is moved over a conductor, a voltage will be induced within the conductor. *Induced* merely means the voltage will be produced within the wire by the magnetic flux. The wire and magnet need not touch physically.

An alternating current generator, as shown in Figure 1–10, uses this same inductive principle. By moving a coil of wire (rotor) through a magnetic field of a permanent magnet, it converts mechanical energy into electrical energy.

To make the operation easier to follow, the rotor has been divided into a dark and light half. When the dark side of the rotor is first rotated through the magnetic field at position A, it is moving parallel with the magnetic lines of force from the permanent magnet. As a result, no voltage is generated. Voltage begins to rise, and in the positive direction, as the rotor continues to rotate and cuts across more magnetic lines of force, until it approaches position B. Voltage is most intense at this position, because the rotor is cutting directly across the greatest concentration of magnetic lines of force. As the rotor continues toward posi-

FIGURE 1–10
Basic alternating
current generator

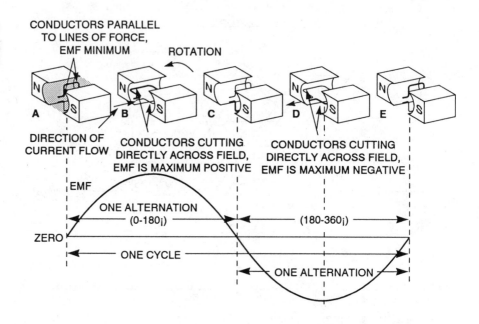

tion C, fewer lines of magnetic force are cut, until voltage again drops to zero. At this point the rotor has turned 180° from its starting position.

As the rotor continues to turn, voltage will again intensify, but now in the opposite, or negative, direction and will reach its maximum at position D. Finally, as the rotor turns a full 360° from where it first started, to position E, voltage will return to zero. The rotor has completed one full cycle.

Figure 1–11 shows a high-speed chart with a typical AC voltage plotted over a one-minute period. It can be seen that the cycle described above occurs 60 times each second. Each cycle takes only 1/60th of a second, or 0.0167 seconds.

Figure 1–12 shows just one cycle, to make it easier to study. Notice the terms identifying the positions on the cycle: peak voltage, effective voltage, and average voltage.

FIGURE 1–11
60 HZ AC

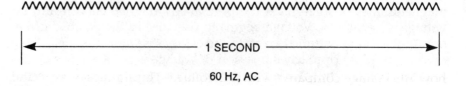

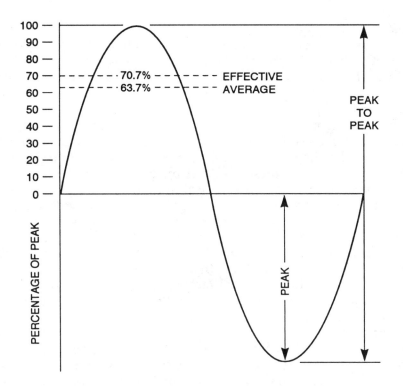

FIGURE 1–12
AC sine wave

Peak voltage is the maximum point the voltage reaches on both the positive and negative portion of each cycle. In the description of the generator above, it will occur at positions B and D, when the rotor cuts directly across the maximum number of magnetic lines of force. Designers of electronic equipment use peak voltage quite often. For example, in deciding the voltage rating of capacitors to be used in a circuit, peak voltage must be considered. Special voltmeters are available to read peak voltage or peak-to-peak voltage. It can also be read with an oscilloscope. However, for most electrical work, it is not a practical measurement.

To calculate peak voltage, take a voltage reading on a regular voltmeter. For example, voltage from a normal house wall outlet will be approximately 120 volts. Multiply 120 × 1.414 to come up with a peak voltage of 169.7 volts. Peak-to-peak voltage would be two times peak voltage, or 339.4 volts.

Effective voltage is called rms or root-mean-square voltage. Near the end of the last century, engineers had to develop a system to measure how AC voltage compared with DC voltage. The simplest way found was to put both AC and DC voltage through resistors. In both cases the resistor gave off heat. Experiments proved that AC peak voltage of 169.7 volts did not produce as much heat as 169.7 DC volts. In fact, the peak AC voltage only produced 70.7 percent as much heat as 169.7 volts DC did. That is, a peak voltage of 169.7 volts AC was only as effective as 120 volts DC (169.7 × .707 = 120 effective volts). A look at the sine wave in Figure 1–12, page 15, will show why. While DC voltage is steady, neither varying up or down, AC voltage has many levels. It starts from zero, climbs in a positive direction to a maximum point, then returns to zero before heading in a negative direction to a maximum point and then to zero. Over that one cycle, there are many levels of intensity.

Root-mean-square is a mathematical method of taking many points along the sine wave and calculating instantaneous voltage. These points are squared, added together, averaged, and the square root extracted. While we won't go through the mathematics here, it turns out that the rms value of 169.7 peak AC volts equals 120 volts (169.7 × .707). The rms, or effective voltage, is the most commonly used voltage in electrical work. While most voltmeter scales are calibrated to give a reading in terms of rms, or effective volts, they do not read true rms voltage. Instead they read average voltage.

If a number of voltage points are taken along one half of a sine wave, added together, and then divided by the number of points taken, *average voltage* is indicated. Average voltage is equal to 63.7 percent of peak voltage. Using the same peak voltage numbers as above, average voltage is 169.7 peak volts × 0.637 = 108.1 volts. Many voltmeters and volt-ohm milliammeters actually read out in average values but have scales calibrated to read in equivalent rms.

Most electrical workers have total respect for voltages of 240, 480 volts or higher. There is no doubt in their minds that these voltages are often deadly if bare skin makes contact with them, but 120 volts?

Many can tell of times they have accidentally brushed against 120 volts and barely felt a tingle. This is also true—sometimes.

The fact is that most deaths in the electrical industry are caused by 120 volt circuits.

This is why the Department of Labor, under the Occupational Safety and Health Administration (OSHA), has issued rules for personnel exposed to possible contact with voltages 50 volts or higher. OSHA says if employees are in a position to make contact with uninsulated live circuits of 50 or more volts, they must wear protective equipment, such as rubber gloves. They must also receive instructions about how to work safely in that area. Chapter 4 will go into greater detail on the use of safety equipment.

> **WARNING**
> *In dealing with test instruments and electrical measurements, you are talking about a commodity—electricity—that most of us take for granted. It has been a loyal servant in working for us and providing entertainment, and it is easy to overlook the fact that, under certain conditions, it can also be extremely dangerous.*

SUMMARY ████████████████████████████

Electricity is the flow of electrons. In order to get electrons to flow in a complete circuit, some sort of pressure must be applied. Called volts, this pressure can be steady and unchanging, such as DC, or it can change direction, such as AC.

Measuring instrument circuits provide a complete path that ends up going through a meter. To measure properly, they must be designed so voltage, current, and ohms are balanced to provide sufficient current to deflect the meter up scale. Test equipment designers use both series and parallel circuits to do so. Because AC voltage cycles, three types of values—peak, rms, and average—must be considered, depending on how the measurements will be used.

Safety is an important factor at all times, but especially when a technician is working on energized circuits. OSHA has made it quite clear at what voltage limits it is unsafe to work without protection. Keep that in mind whenever using an instrument to take electrical readings.

REVIEW QUESTIONS ████████████████████████

1. In order to have electricity, what must flow?

2. Since the filament of a bulb is made of wire, why doesn't it act like a short circuit when wired into a circuit?

3. In a series circuit, must the current be calculated for each resistor?

4. Is the voltage drop across each leg of a parallel circuit different?

5. Why is the total resistance in a parallel circuit equal to or less than the smallest resistor?

6. Why is peak voltage in AC not equal to the same voltage in DC?

7. If peak voltage is the highest value reached during each AC cycle, why do most meter scales read out in rms voltage?

8. The power formula E x I can be used for both DC and AC when operating what electrical device?

9. Watts means work performed. What work is performed when AC or DC voltage is put through a resistor?

10. What is the lowest voltage that OSHA says a worker can be exposed to without protective equipment?

Introduction to
Analog Test
Equipment
Measuring Circuits   2

INTRODUCTION

Although most test equipment purchased today is digital, few maintenance departments can afford to scrap all their existing analog equipment. Therefore Chapter 2 will review the more common analog
measuring instruments and circuits. As when using any piece of test
equipment, the technician should first consult the appropriate instruction manual. These publications contain detailed and specific information on the particular equipment to be used. Often such manuals are
missing or not available. It is not intended for this textbook to take the
place of a manufacturer's instruction manual. Nevertheless, it will give
the student some idea of the safe and efficient use of test equipment
found in many electrical maintenance shops.

OBJECTIVES

After studying this chapter, the student should be able to:

- *List three basic types of analog meter movements.*
- *Describe how meters measure DC circuits.*
- *Show how an ohmmeter circuit works.*
- *Describe how meters measure AC circuits.*
- *Understand the problems when working with high-voltage power
  circuits.*

ANALOG VERSUS DIGITAL METERS

Analog is derived from the word to analyze. In the electrical world, it means to measure the magnitude of some quantity we cannot see—volts, amperes, or ohms—and display the results in some physical form we can see. An example might be the pointer of an electro-mechanical voltmeter reading the voltage from a wall outlet. As the voltage rises or falls, the meter pointer also rises or falls. By reading where the pointer is positioned over a calibrated scale, the magnitude of the voltage measured can be read.

Digits means numbers. In order for a digital meter to measure voltage, for example, circuitry within the meter must change the magnitude of volts to a form that can be counted electronically. Then the meter will count and display the numbers representing the voltage measured. Digital meters will be covered later in the text.

ANALOG METER MOVEMENTS

A number of meter elements are used in analog meters. All use a variation of the motor action principle as shown in Figure 2–1. Two fluxes interact to deflect a pointer up scale. However each element has its own method of operation.

FIGURE 2–1
Motor action

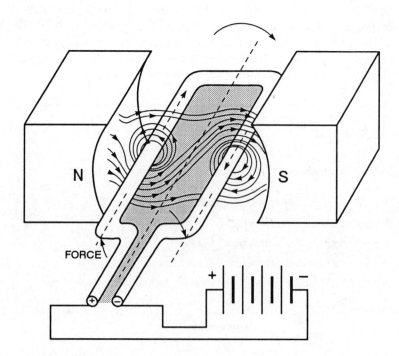

D'Arsonval Movement

Most analog measuring instruments for servicing electrical equipment use a stationary, permanent, magnet-moving, coil meter (PMMC). This type of movement is called the D'Arsonval movement. A current-carrying coil is supported by jeweled bearings between the poles of a U-shaped permanent magnet, as shown in Figure 2–2. A soft-iron stationary member mounted between the poles of the magnet completes the magnetic circuit and concentrates the field through the coil. The length of the conductor in the coil and the strength of the field between the poles of the magnet is fixed. Therefore, any change in current causes a proportionate change in the force acting on the coil.

Spiral hairsprings are physically located on the front and back of the element and connected to the coil. Conductors connect the hairsprings with the outside terminals of the meter. Current flows into one of the hairsprings through the coil, and out of the other hairspring. Magnetic

FIGURE 2–2
D'Arsonal movement

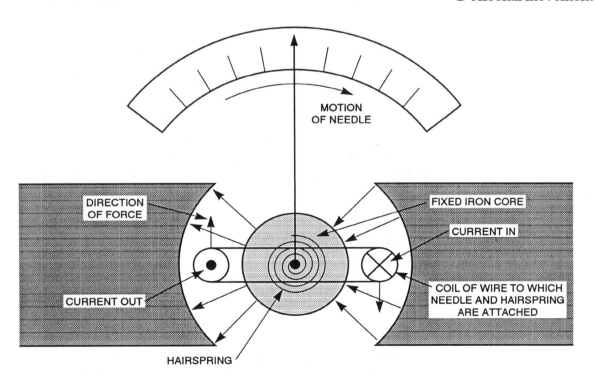

fields within the movable coil work against the permanent field of the magnet to deflect the needle up scale and wind up the hairspring. When the current through the coil is interrupted, the restraining force of the spiral springs returns the pointer to the normal, or zero, position. The front spiral spring is also used to zero the pointer. Some pointers have balance arms with movable counterweights that can be adjusted to insure that the meter will read correctly when in either the vertical or horizontal position. As protection against physical damage to the pointer from overloads or incorrect polarity, mechanical stops are located at the beginning and end of pointer travel.

The D'Arsonval meter movement relies on a magnetic field developed in the moving coil, working against the permanent magnet, to deflect the pointer up scale. As a result, a steady rate of magnetic flux must flow through the coil, which can only be provided by DC current. Although the movement is polarity sensitive, meters are available with a center zero position that can deflect to the left or right of center, to show both magnitude and direction of polarity.

If AC current is fed to the movable coil, the pointer will try to deflect up scale, but each time the AC sine wave goes through zero in its cycle the pointer will drop back to zero. Since the pointer cannot react quickly enough to these changes in direction, the pointer will simply vibrate in position and will not give an up scale indication.

Figure 2–3A shows the typical horseshoe magnet used on this type of meter element. Alnico magnets (aluminum, nickel, and cobalt) cast in a rectangular slug are shown in Figure 2–3B. Flux from this type of magnet divides into the two sides of the soft iron rings surrounding the slug. These rings also act as shields to protect the magnet assembly from outside magnetic disturbances.

In better meters, the pointer quickly settles down without overshooting or oscillating around the correct reading. This is called damping and is accomplished in many D'Arsonval movements by winding the coil on an aluminum bobbin. As the bobbin spins in the magnetic field, an electromotive force (emf) is induced in it, much like in an AC generator when the loop of wire cuts through the magnetic lines of force. Induced currents flow in the bobbin in such a direction as to oppose the motion of the bobbin, and the pointer quickly comes to rest in the final position.

Taut Band

Another member of the D'Arsonval family is the taut band meter element, shown in Figure 2–4. Instead of jeweled bearings and coil springs, these meter elements have thin, flat, phosphor ribbons that

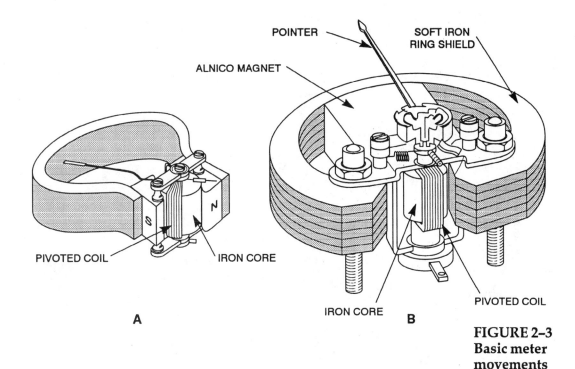

POINTER

SOFT IRON
RING SHIELD

ALNICO MAGNET

PIVOTED COIL

IRON CORE

A

IRON CORE

B

PIVOTED COIL

**FIGURE 2–3
Basic meter
movements**

provide the conducting path for the current between the current under test and the movable coil. They also provide the restraining force for the movable coil. As current enters the movable coil, a deflection occurs between the coil and flux field and the permanent magnet flux field. The ribbons supporting the coil twist through an angle proportional to this force. When the twisting force equals the magnetic force, the pointer stabilizes and a measurement of current is obtained. Once the driving force of the coil current is removed, the force stored in the twisted ribbons returns the coil to its zero position. Because there are no jeweled bearings in the unit, repeat readings are more likely, and damage from rough usage is less of a problem.

Taut band meter elements are merely a variation of the D'Arsonval movement. As such they only function on DC current and are also polarity sensitive.

ELECTRODYNAMOMETER

The electrodynamometer differs from the D'Arsonval movement in that no permanent magnet is used. Instead, two fixed coils produce the

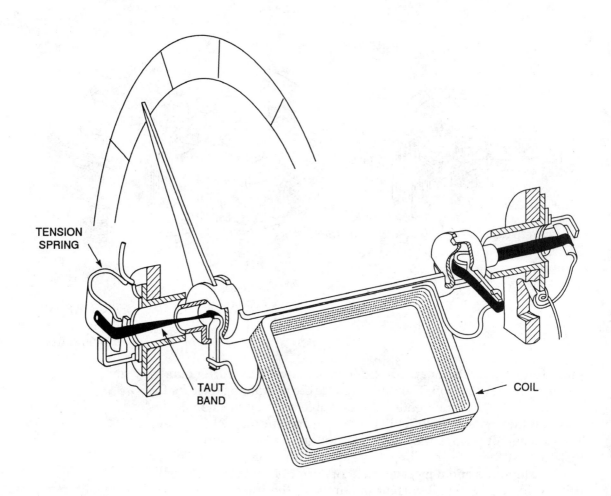

**FIGURE 2–4
Taut band
movement.
(Courtesy of Triplett
Corporation—Warren
Hess, President)**

permanent magnetic fields, and two movable coils use this same magnetic field to create pointer deflection, as shown in Figure 2–5.

When used as an ammeter, all coils, both permanent and movable, are in series with each other. In the meter, however, the two fixed coils are positioned so that both turn around a common axis with a space between them. The two movable coils are mounted on a pivot between the fixed coils, but have their axis at right angles to the axis of the

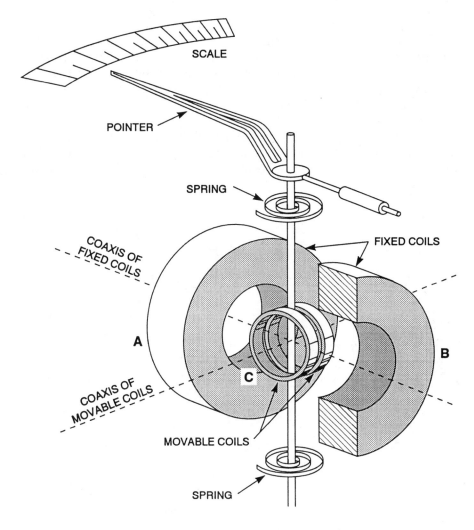

FIGURE 2–5
Electrodynamometer

permanent coils. When current enters the movable coil they try to align their axis with those of the permanent coil and deflect, together with the attached pointer, up scale.

The central shaft, on which the movable coils are mounted, is restrained by spiral springs that hold the pointer at zero when no current is flowing through the coil and serve as conductors for delivering current to the movable coils. Since these conducting springs are very small, the meter cannot carry a very heavy current.

When the electrodynamometer is used as a voltmeter, no problem is encountered, because the current is only 0.1 ampere, and can be brought in and out of the moving coil through the springs. Internal connections and construction of a voltmeter are shown in Figure 2–6A. Fixed coils "a" and "b" are wound with fine wire and are connected directly in series with the movable coil "c" and the series current-limiter resistance.

For ammeter applications, however, a special type of construction must be used, because the large currents that flow through the meter cannot be carried through the moving coils. In Figure 2–6B, stationary

FIGURE 2–6
Circuits

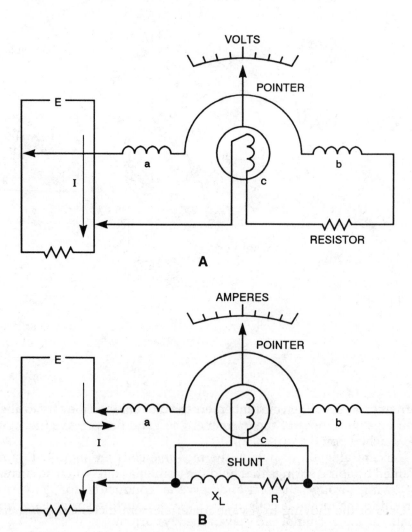

coils "a" and "b" are generally wound of heavier wire, to carry up to 5 amperes. In parallel with the moving coils is an inductive shunt, which permits only a small part of the total current to flow through the moving coil. Current through the moving coil is directly proportional to the total current through the instrument, as the shunt has the same ratio of reactance to resistance as the moving coil. Therefore the instrument will be reasonably correct at all frequencies that it is designed to be used.

The electrodynamometer is mechanically damped by means of aluminum vanes that move in enclosed air chambers. Although electrodynamometers are very accurate meters, they do not have the sensitivity of the D'Arsonval type meter and are not widely used outside the laboratory. One of the more common uses of the electrodynamometer is for analog phase-angle meters and watt meters.

Moving Iron-Vane Meter

The moving iron-vane meter is another basic type of meter element. Unlike the D'Arsonval type meter, which employs permanent magnets, the moving iron-vane meter depends on induced magnetism for its operation. It employs the principle of repulsion between two concentric iron vanes, one fixed and one movable, placed inside a coil, as shown in Figure 2–7. A pointer is attached to the movable vane.

When current flows through the coil, the two iron vanes both become magnetized with north poles at their upper ends and south poles at their lower ends. Because like poles repel, the two iron vanes move away from each other and against the force exerted by the springs. Repulsion is always in the same direction, regardless of the direction of current through the coil, so the moving iron-vane instrument operates on either DC or AC circuits. Because magnetic flux from the magnetic coil is induced in both iron vanes, the pointer travel up scale is the result of I^2 current, and the scale plate is not linear. When used on AC circuits, the instrument has an accuracy of 0.5 percent. Some loss of accuracy occurs when used on DC.

Mechanical damping is obtained by using an aluminum vane attached to the shaft (not shown in the figure), so as the shaft moves, the vane moves in a restricted air space.

Uses of the moving-iron-vane meter element include panel meters and some clamp-on ammeters.

FIGURE 2–7
Moving iron-vane
meter

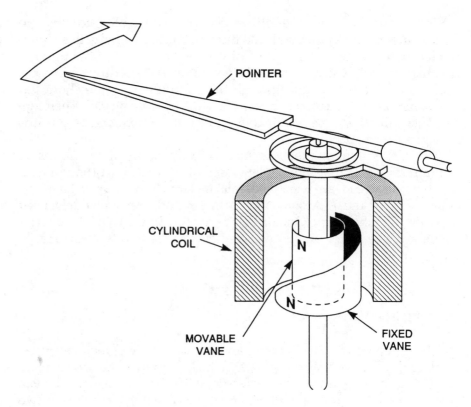

THERMOCOUPLE

In this chapter we have covered analog metering elements. Because the thermal method is an analog system for measuring current flow, this is an appropriate place to consider its operation.

The thermal method uses a heater, and a device called a thermocouple, in which the ends of two dissimilar metals are welded together, as shown in Figure 2–8. If this junction is heated, a DC voltage is developed across the two open ends. The voltage developed depends on the metals used and the difference in temperature between the heated junction and the open ends.

In the thermocouple, the junction is heated electrically by current flow through a heater element made of a resistant material. It does not matter whether the current is AC or DC because the heating effect is not dependent on current direction. The maximum current measured depends on the current rating of the heater, the heat the thermocouple can withstand without damage, and the current rating of the meter used

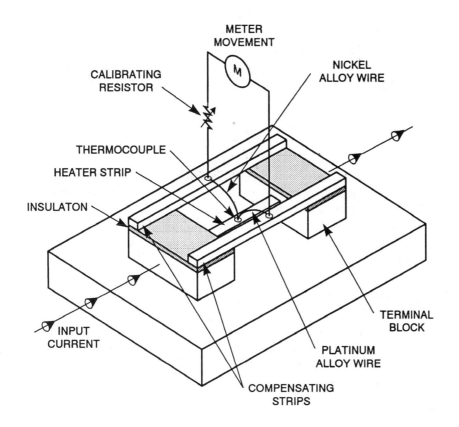

FIGURE 2–8
Thermocouple

with the thermocouple. Voltage may also be measured if a suitable resistor is placed in series with the heater.

Input current, by way of the terminal blocks, flows through the heater strip, where two dissimilar wires are welded to form the junction of the thermocouple. The open ends of these wires are connected to the center of the compensating strips. The compensating strips, insulated thermally and electrically from the terminal blocks, radiate heat so the open ends of the thermocouple wires are much cooler than the junction end. This permits a higher voltage to be developed.

Heat produced by the flow of line current through the heater strip is proportional to the square of the heating current ($P = I^2R$). Since voltage appearing across the two open terminals is proportional to the temperature, movement of the meter element connected across these terminals is proportional to the square of current flowing through the heater element. The scale of the meter is crowded near the zero end and progressively expands near the maximum end, as shown in Figure 2–9. Consequently, readings are less accurate at the low end. It is desirable

FIGURE 2–9
Non-linear scale for
thermocouple meter

to choose a range in which the deflection will extend at least to the wider portion of the scale. The meter used has a low resistance to match the thermocouple, and it must deflect full scale when rated current flows through the heater. Since the resistance must be low and the sensitivity high, the moving element must be light, so D'Arsonval movements are often used.

MEASURING CIRCUITS

So far various meter elements and thermocouple devices have been covered. Now it is time to see how these units are used in test instruments to read volts, amperes, and ohms.

DC Voltmeters

Voltmeters, especially multimeters to be covered later in this chapter, have ranges of 1,000 volts and above. The insulating material used on test leads is usually rated at 600 volts. Never rely on insulation over alligator clips or wire to protect you from energized conductors. Turn the circuit off. Make your connections; then turn it back on.

In discussing how various analog measuring elements operate, it was always in terms of current and magnetic flux. What happens if voltage needs to be measured? Ohm's Law shows that current cannot flow without voltage. If that is the case, it seems that by putting the DC voltage to be measured through some resistance a current would be produced. Figure 2–10 shows such a simplified circuit for reading DC voltage.

In order for a voltmeter to give accurate measurements, it must not add the burden of its own measuring circuit to that of the one being measured. How well it handles this task is called sensitivity. Sensitivity, measured in ohms per volt, equals the amount of internal current needed to deflect the meter mechanism full scale. This rating is often marked on the nameplate of the meter. Typical current values for full-scale deflection range from 10 microamperes to 100 milliamperes for

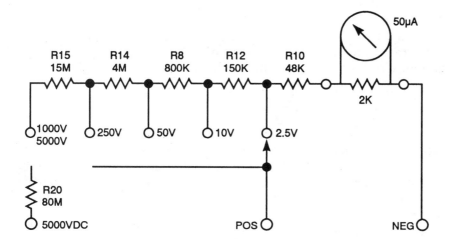

FIGURE 2–10
DC voltmeter

inexpensive units. Ohms per volt can be calculated by taking the reciprocal of the full-scale current rating of the meter. For example: the meter element shown in Figure 2–10 has a full-scale current rating of 50 microamperes (.000050). Ohms per volt equals:

$$\frac{1}{.000050} = 20,000 \text{ Ohms per Volt}$$

In order to measure DC voltage, resistors must be put in series with the measuring meter element. Called multiplier resistors, their value will be a simple function of Ohm's Law.

To measure one volt, 20,000 ohms of resistance must be in the circuit. Because the lowest voltage range shown in the simple circuit in Figure 2–10 is 2.5 volts, the resistance must be:

$$\frac{2.5}{.00005} = 50,000 \text{ Ohms}$$

Since the meter already has an internal resistance of 2,000 ohms, a multiplier resistance of 48,000 must be added so the meter will read full-scale at 2.5 volts.

DC voltmeters are connected in parallel, with polarity observed. The pointer will indicate from left to right. When reading an unknown voltage, it is always best to start with the highest range. Select lower ranges until the pointer indicates a reading of at least 2/3 of full scale.

High-voltage probes are available for reading voltages of TV picture tube circuits or the spark plugs of cars. Shown in Figure 2–11, these

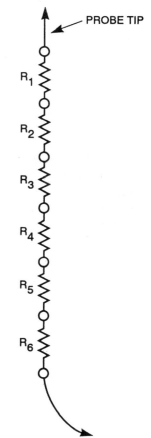

FIGURE 2–11
High-voltage probe

devices are a series of dropping resistors that allow a controlled voltage
to reach the meter.

All these probes are designed to handle the high-voltage but low-
ampere circuits of the TV or automobile.

DC Ammeters

Most DC ammeters are in-line devices and are polarity sensitive: that is,
the meter is put in series with the load being measured. Because they
are polarity sensitive, care must be taken to connect the plus and minus
leads to the proper test points. If polarity is reversed, the needle will
deflect backward and may damage the movement. Meters should always be connected so the electron flow will be into the negative terminal and out of the positive terminal. Since the current must pass
through both the meter and the device, the ammeter must be connected
in series with the device that is being tested. The meter used in the
circuit shown in Figure 2–12 can handle up to 50 microamperes without
additional circuitry. So to create a multi-range ammeter, internal resistors must be connected in parallel with the meter measuring circuit.

**FIGURE 2–12
Ammeter circuit**

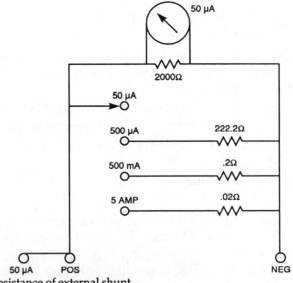

To calculate resistance of external shunt.

$$\text{Resistance of shunt} = \frac{\text{I of Meter} \times \text{R of Meter}}{(\text{I Total} - \text{I Meter})}$$

Total current = 100 Amperes

$$\text{Rshunt} = \frac{0.00005 \times 2000}{(100 - 0.00005)} = \frac{0.1}{99.99995} = .001\Omega$$

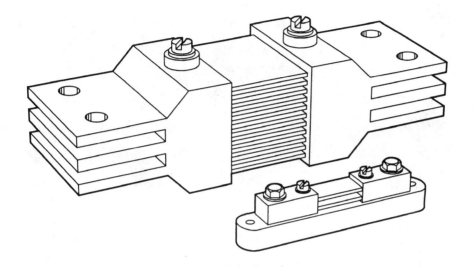

**FIGURE 2–13
Ammeter shunts**

Resistor values are calculated so all measured current over 50 microamperes is shunted around the meter.

Should it be necessary to measure larger currents than internal resistors can handle, external shunts are available. Shunts are usually made of manganin, an alloy having almost zero temperature coefficient of resistance. The ends of the shunt strips are embedded in heavy copper blocks, for ease of connection, and meter coil leads are attached to the blocks, as shown in Figure 2–13. To ensure accurate readings, the meter leads for a particular ammeter should not be used interchangeably with other meters of a different range. Slight changes in lead length and size can vary the resistance of the meter circuit and its current, and cause incorrect meter readings.

Calculating an external shunt is similar to the procedure used above for calculating the internal ranges.

Clamp-on DC Ammeters

There are also DC hook-on ammeters manufactured with a split core of magnetic material. This instrument has an opening in the bottom section for the insertion of different DC meter movements, depending on the range needed. The instrument is similar to other hook-on instruments, but the meter movements are encased and detachable. The movements may differ in size as the type and marking of the scales change. This type of instrument has only limited application.

**FIGURE 2–14
AC/DC Hall effect
probe. (Courtesy of
John Fluke
Manufacturing
Company, Inc.)**

Also available are clamp-on devices that will work with either AC or DC current. Using what is called a Hall effect, these battery-operated devices have jaws that contain a semi-conducting material and output a DC voltage in proportion to the current measured. Figure 2–14 shows such a device, which can be used with any DC reading voltmeter.

OHMMETERS

Ohmmeters are widely used to measure resistance and check the continuity of electrical circuits and electronic devices. For most multimeters, resistance ranges are up to and including megohms.

An ohmmeter circuit consists of a DC milliammeter and a battery to supply potential to the measuring circuit. A number of resistors are included, with one of them variable, to adjust current flow in the circuit as battery strength weakens. Some meters include a second battery to ensure sufficient current flow in high-resistance measuring ranges. A simple ohmmeter circuit is shown in Figure 2–15. Rarely does an ohmmeter have two batteries for two ranges; however, the circuit does show the principle involved in a high-range, low-range meter.

Connection

Deflection of an ohmmeter pointer is controlled by the amount of DC current passing through the moving coil. When measuring an unknown resistance, the ohmmeter test leads are first shorted together. This results in zero resistance being measured and maximum current flow in the measuring circuit, so the pointer deflects up scale to the

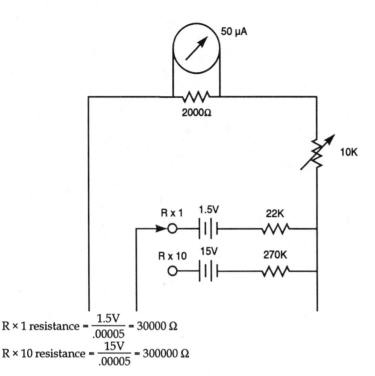

FIGURE 2–15
Ohmmeter circuit

$$R \times 1 \text{ resistance} = \frac{1.5V}{.00005} = 30000 \ \Omega$$

$$R \times 10 \text{ resistance} = \frac{15V}{.00005} = 300000 \ \Omega$$

right. The variable resistor (rheostat) is adjusted until the pointer of the meter comes to rest exactly on the zero mark.

Notice again that zero resistance being measured means maximum current flow in the meter element. That is why the zero position on the ohm scale is exactly opposite from the zero position on the volt and ammeter scale, as shown in Figure 2–16. When the test leads are separated, and no current flows, spring tension returns the pointer to the left side of the scale. Once the ohmmeter is adjusted for zero reading, it is ready to measure resistance. Any resistance placed between the probes is in series with the ohmmeter circuit. More resistance means less current flow, and the pointer will drop off of full scale and indicate the resistor value.

Measurement

Resistance measurements must always be performed on de-energized circuits, because voltage applied across the meter could cause damage to the meter movement or meter resistors.

FIGURE 2–16
Ohmmeter scale.
(Courtesy Simpson
Electric Company)

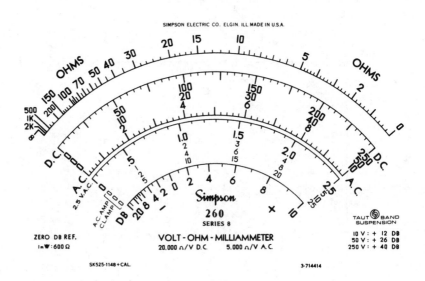

The test leads of the ohmmeter are connected so the meter and battery are in series with the circuit to be measured, as in Figure 2–15. This causes the current produced by the meter internal battery to flow through the circuit being tested. Because the meter has been preadjusted (or zeroed), the amount of coil movement now depends solely on the resistance measured to raise the total series resistance, decrease the current, and thus decrease the pointer deflection.

If R_1 or R_2, or both, were replaced with a resistor having a larger ohmic value, the current flow in the moving coil of the meter would be decreased even more. Deflection of the pointer would be further decreased, and would read on the scale as higher circuit resistance. Response of the moving coil is proportional to the current flow, while the scale readings of the meter, in ohms, are inversely proportional to current flow in the moving coil.

Resistance to be measured may vary from only a few ohms to megohms (million ohms) in some meters. To enable the meter to show the value being measured, with the least error, most incorporate a scale multiplication feature. For example, a typical meter has test positions marked R x 1, R x 10, R x 100, and R x 1,000. A switch selects the multiplication scale desired. The range to be used in measuring any particular unknown resistance (R_{10} in Figure 2–15) depends on the approximate ohmic value for R_X, the unknown resistance.

Notice the ohmmeter scale is not linear. Ohmic values on the left-hand side of scale are extremely crowded; therefore the most accurate readings are from the center to the right edge of the ohmmeter scale.

AC VOLTMETERS

A voltmeter indicates the potential difference between two points in a circuit. Voltmeters are designed to measure DC or AC voltage. Remember that DC voltage is steady state; it does not change in intensity. AC voltage changes direction twice in each cycle. Because of this change in direction from the positive to zero to negative to zero, the measuring meter element must be able to respond to the AC sine wave. This limits choices to the electrodynamometer, iron vane, or thermocouple method. If AC is to be measured with a meter using a variation of the D'Arsonval movement, then the AC sine wave must be changed to a DC voltage first. This is done using a device called a rectifier.

RECTIFIER METHODS

A rectifier is a device that opposes current flow through it in one direction, but not in the other. It is a directional conductor, and is used mostly for converting alternating current into a direct current. Early models of electrical test equipment used metallic rectifier elements of two basic types: copper oxide and selenium. Although both still exist today, these types of rectifiers have been replaced by the germanium and silicon diode in test equipment applications. These rectifiers will be covered in this chapter.

A metallic rectifier element has a good conductor and a semiconductor (material of high resistivity) separated by a thin, insulating, barrier layer. The flow of forward current through a cell has a flow of electrons from the good conductor, across the barrier layer, and through the semiconductor.

METALLIC RECTIFIERS

Metallic rectifier cells are usually made as plates, circular or square in shape, with a hole in the center. Many cells, with the necessary terminals, spacers, and washers, are assembled on an insulated stud passing through their center holes. This assembly is called a rectifier stack. Some rectifier stacks contain fins, which are used to keep the rectifier

SCHEMATIC SYMBOL

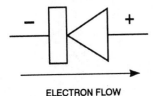

ELECTRON FLOW

FIGURE 2–17
Rectifier symbol

from overheating; they afford a large surface area for conducting away the heat. Cells and stacks can be connected in series or parallel, with proper polarities, to obtain the required voltage, current ratings, and circuit connections for specific applications.

Two types of metallic rectifiers are used: (1) a thin film of copper oxide and copper, and (2) selenium (either iron or aluminum). Rectifier units are represented by the symbol shown in Figure 2–17, with the arrowhead in the symbol pointing against the direction of electron flow.

Figure 2–18 shows a simple AC circuit that uses a rectifier to produce a series of DC voltage pulses as its output.

Copper-Oxide Rectifier

In the copper-oxide rectifier, the oxide is formed on the copper disk before the rectifier unit is assembled, as shown in Figure 2–19A. In this type of rectifier, electrons flow more readily from the copper to the oxide than from the oxide to the copper. External electrical connections can be made by connecting terminal lugs between the left pressure plate and the copper, and between the right pressure plate and the lead washer.

FIGURE 2–18
Waveform circuits

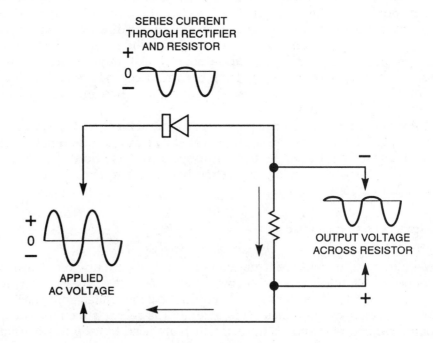

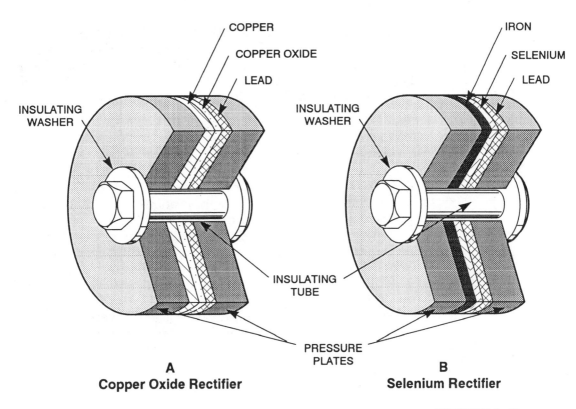

A
Copper Oxide Rectifier

B
Selenium Rectifier

FIGURE 2–19
Metallic rectifier

For the rectifier to function properly, the oxide coating must be very thin. Because rectifiers are polarity sensitive, each unit can only stand a low voltage, called inverse voltage, from the wrong direction. Rectifiers designed for moderate and high powered applications consist of many of these individual units mounted in series on a single support. The lead washer enables uniform pressure to be applied to the units, so the internal resistance may be reduced. When the units are connected in series, they normally present a somewhat high resistance to the current flow. Resultant heat developed in the resistance must be removed if the rectifier is to operate satisfactorily. Useful life of the unit is extended by keeping the temperature below 140° F. The efficiency of this type of rectifier is generally between 60 and 70 percent.

Selenium Rectifiers

Selenium rectifiers, shown in Figure 2–19B, function in much the same manner as copper-oxide rectifiers. Such a rectifier is made up of an iron disk coated with a thin layer of selenium. In this type of rectifier, the electrons flow more easily from the selenium to the iron than from the iron to the selenium. Selenium rectifiers may be operated at a much higher temperature than a copper-oxide rectifier of similar rating with efficiency between 65 and 85 percent, depending on the circuit and the loading.

Diodes

Rectifiers can be used both as half-wave rectifiers and in full-wave and bridge circuits. In each of these applications, the action of the metallic rectifier is similar to that of a diode.

Many types of semiconductor rectifier diodes are available, with germanium and silicon being used extensively. These rectifiers vary from the size of a pinhead used in miniature circuitry, to large, 500-ampere diode rectifiers used in power supplies. Silicon diode rectifiers that are one inch long and one inch in diameter will supply a direct current of 50 amperes (peak), and have a peak-inverse voltage rating of 60 volts. These units operate in much the same manner as the metallic rectifier and are used in similar applications.

Use of Solid-State Rectifier Circuits

It is possible to couple a D'Arsonval direct-current type instrument and a rectifier to measure AC quantities. Rectifiers are usually semiconductors arranged in a bridge circuit, as shown in Figure 2–20. By using a bridge circuit, current always flows through the meter in one direction. When the voltage being measured has a waveform as shown in Figure 2–20, the path of current flow will be from the lower input terminal through rectifier No. 3, through the instrument, and then through rectifier No. 2, thus completing its path back to the source's upper terminal. The next half cycle of the input voltage (indicated by dotted sine wave) will cause the current to pass through rectifier No. 1, through the instrument, and through rectifier No. 4, completing its path back to the source.

This type of instrument is characterized by errors due to waveform and frequency, and allowances must be made according to data furnished by the manufacturer. Corrective networks can be added to the instrument that will make it practically error free up to 100 kHz. Instruments of this type only require one milliampere of current from the line for full scale deflection, and are widely used for low-range AC voltmeters.

Some "aging" of the rectifier, with a corresponding change in the calibration of the instrument, does occur, and such instruments must be

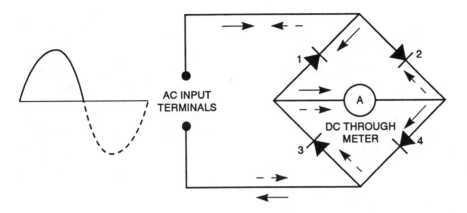

FIGURE 2–20
Full wave rectifier

recalibrated from time to time. Rectifier-type instruments read the average value of the AC quantity. Yet, because the effective or root-mean-square (rms) values are more useful, AC meters are generally calibrated to read rms values (1.11 times the average of the instantaneous values).

Connection

When the voltmeter is connected across a circuit, it shunts the circuit. If the voltmeter has low resistance, an appreciable amount of current will flow through the meter. The effective resistance of the circuit will be lowered, and voltage reading will be lowered due to the increase in current flow.

When voltage measurements are made in high-resistance circuits, it is necessary to use a high-resistance voltmeter to prevent the shunting action of the meter. The effect is less noticeable in low-resistance circuits, because the shunting effect is less.

Voltage-measuring instruments are connected across (in parallel with) a circuit. If the approximate value of the voltage to be measured is not known, it is best to start with the highest range of the voltmeter and progressively lower the range until a suitable reading is obtained. This is done with the RANGE switch. Care must be taken not to over range the instrument (go to a range lower than circuit voltage).

Measurement

The basic 50-microampere D'Arsonval movement used for the ammeter can measure voltage if a high resistance is placed in series with the moving coil of the meter. Because the value of the resistor is constant, any change in current through the coil is proportional to the voltage measured. Using this principle, the instrument indicates voltage. The meter is reading voltage, although activated by current. In practice, the voltage ranges of the instrument are established by different values of multiplying resistors. For low-range instruments, these are made from temperature coefficient resistance wire wound either on spools or card frames inside the meter. For higher voltage ranges, the series resistance may be externally connected to the meter movement using high precision resistors called multipliers.

AC AMMETERS

AC ammeters are of two basic types: in-line, where the meter becomes a part of the circuit measured, and clamp-on types, where readings are taken by induction.

In-line Ammeter

Application. An ammeter is used to measure current flow in an electrical device or circuit. The range of the basic movement is very small (milliamperes). For the majority of applications, a shunt is internally or externally added to the basic movement to extend its range. Figure 2–21 is a photo of an in-line ammeter.

Connection. If an ammeter were connected in parallel across a constant-potential source in the manner a voltmeter is used, the internal shunt would become a short circuit and the meter would burn out. Ammeters have a very low resistance; therefore, another resistance component must be connected in parallel with it to limit the current to a safe level.

If the approximate value of current in a circuit is not known, it is best to start with the highest range of the ammeter and switch to progressively lower ranges, until a suitable reading is obtained.

To measure current in an AC circuit correctly, the meter should not interfere with the current being measured. The small size of the wire with which an ammeter's movable coil is wound places severe limits on the current that may be passed directly through the coil. Basic D'Arsonval movement can only measure very small currents, such as microamperes (10^{-6} amperes) or milliamperes (10^{-3} amperes).

To measure larger currents, a current transformer must be used with the meter movement. Transformers are usually rated as primary current/secondary current as 400/5. Figure 2–22 shows such a test set up.

FIGURE 2–21
In-line ammeter.
(Courtesy of AVO
Multi-Amp
Corporation)

One word of caution about using current transformers. As current flows through the primary of a current transformer, it sets up a magnetic field in the core and is induced in the secondary. As long as a path exists across the terminals of the secondary, such as a shorting bar, or the coil of an ammeter, the secondary current will be at a ratio equal to the current rating of the CT. That is, if 400 amperes are flowing through the primary, as shown in Figure 2–22, then 5 amperes will flow out of the secondary. If, however, a meter lead is suddenly pulled off the secondary coil, all the magnetic flux produced by the primary coil will be diverted to maintain current flowing in the secondary, which is now open circuited. As a result, a high voltage, called crest voltage, develops across the secondary. For a few cycles, this voltage can be as high as the primary voltage the CT is installed on.

Special instrument transformers are available that combine a number of window and in-line ratios all to one 5-ampere secondary. Figure 2–23 shows the layout of such a device.

> **WARNING**
> *A basic rule for using current transformers is never open the secondary of a CT under load.*

FIGURE 2–22
Current transformer

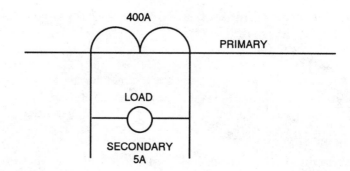

By wrapping one, two, or four windings of a current-carrying conductor through the window, 800, 400, or 200 amperes can be measured.

The normal dial multiplier to change ammeter readings (5 amperes) to primary current for the 800/5 ratio is 160 (800/5 = 160). If two windings of a conductor carrying 400 amperes are wound through the window, the multiplier will be 400/5, or 80. Four windings of a conductor carrying 200 amperes will give a multiplier of 200/5, or 40.

FIGURE 2–23
Multi-tap instrument transformer

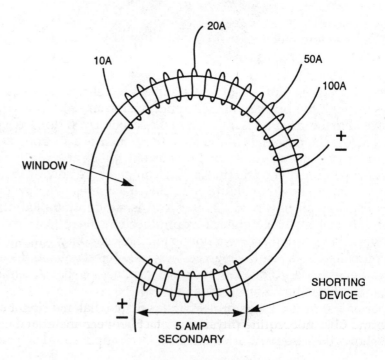

Sound confusing? Think of it this way. A 200-ampere conductor wound four times through the window of the current transformer will impress the same magnetic flux as 800 amperes flowing through the primary. Since the CT secondary will then send 5 amperes to the meter, which is four times the value it should be, the 800/5=160 ratio must be divided by four (160/4) to give the correct multiplier of 40.

When used as an in-line device, the transformer can measure 10, 20, 50, or 100 amperes, all to a 5-ampere secondary. Dial multipliers are found by dividing the input by the output; 100/5 = 20, for example.

When taking readings from a CT, the safest approach is to short out the secondary, using the built-in shorting device. Use spade lugs to install the meter in the secondary, as alligator clips have a nasty habit of slipping off terminals and creating a hazardous situation. Once the CT is installed in the circuit and the meter connected, remove the short. Once readings are taken, short out the CT secondary and replace the original circuit.

AMMETER CONNECTIONS

A single meter movement is used in all ammeter ranges of a particular meter. For example, meters with working ranges of zero to one amperes, zero to five amperes, or zero to ten amperes all use the same galvanometer movement. The designer of the ammeter simply calculates the correct internal resistors required to extend the range of the one-milliampere meter movement to measure the desired amount of current.

Clamp-on Ammeter

Working on the same principles as a current transformer, it is the most often used ammeter.

Application. A clamp-on ammeter has the same application as an in-line ammeter, which is to measure current flow in an electrical circuit. The main difference between the two is in the method of connection.

Clamp-on, or split-core, meters are one of the most commonly used instruments for everyday electrical maintenance. It is a unique member of the ammeter family, because it measures the current flowing in a conductor on a graduated scale by induction. A commonly used type is shown in Figure 2–24. It eliminates the need to break the circuit under test, and does not require physical contact between meter and circuit. This type of instrument is ideally suited to applications where it is either impossible or impractical to open the circuit. The split-core meter is safe, convenient, and time-saving when obtaining alternating-current measurements. It is also capable of reading voltage using the leads provided.

Operation.   Operation of the clamp-on ammeter is similar to that of a current transformer. The conductor being measured is the primary and the jaws of the clamp-on meter the core of the CT. Secondary windings around the core, but within the case, bring AC to a rectifier where it is changed to DC and applied to the D'Arsonval meter circuit, as shown in Figure 2–25. If current in the conductor is so low that little deflection is noticed on the lowest meter scale, the conductor can be wound more than once through the jaws of the device. Just remember that each time you add a winding you must divide the readings by the turns of wire. For example, the pointer reads six amperes on the ten-ampere scale, with three wraps through the clamp-on jaw. Actual ampere is 6/3 or 2 amperes through the conductor.

Accessory clamp-on ammeter adapters are also available from many meter manufacturers. Figure 2–26 shows a unit that will turn any multimeter into a clamp-on ammeter, with switch selectable ranges up to 300 amperes. The output of three volts, AC, on all ranges makes these units easy to use and read. One point to keep in mind is that any error

FIGURE 2–24
Clamp-on ammeter.
(Courtesy of
Amprobe
Instrument®)

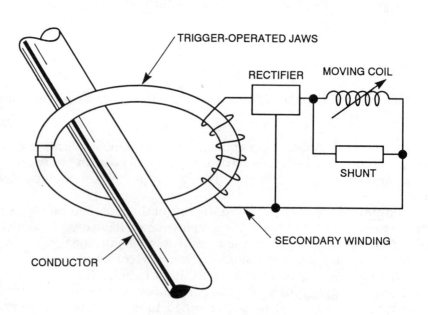

FIGURE 2–25
Typical clamp-on
ammeter circuit.
(Courtesy of
Amprobe
Instrument®)

in the clamp-on device must be added to that of the meter used to come up with the final value read.

To obtain a current measurement with a clamp-on meter, open the jaws of the instrument and place them around the conductor in which current is to be measured. Once around the conductor, allow the jaws to close slowly and securely. The pointer will deflect and indicate the value of current-flow in the conductor.

MULTIMETERS

The multimeter is an instrument that can measure resistance, voltage, or current. It contains one milliammeter with circuitry and graduated scales to indicate the values that can be measured. Figure 2–27 shows one popular type.

The front panel of the multimeter is constructed and labeled so that all functions are self-explanatory. One plug marked -DC ± AC OHMS is common to all functions and ranges. One test lead is always

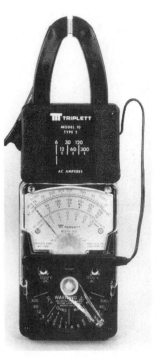

FIGURE 2–26
AC clamp-on adapter. (Courtesy of Triplett Corporation—Warren Hess, President)

FIGURE 2–27
Multimeter. (Courtesy of Simpson Electric Company)

plugged into this common jack, and the remaining lead into the jack marked for the particular function or range desired. A FUNCTION switch selects the proper internal circuitry for the type of measurement desired. The voltage, current, or resistance range is determined by a combination of FUNCTION and RANGE switch settings along with test lead jack positions.

Connection

In connecting test instruments certain procedures should be followed for both safety and accuracy.

Current Measurement. Current-measuring instruments (ammeters) must always be connected in series with a circuit and never in parallel with it.

Voltage Measurement. Voltage-measuring instruments (voltmeters) are connected across (in parallel with) a circuit.

Ohmmeter Measurements. Ohmmeter measurements of resistance must always be performed on de-energized circuits. This prevents the circuit's source voltage from being applied across the meter, which can cause damage to the meter movement and measurement resistor circuits.

Output Jack. If an AC voltage reading is to be made that contains a large DC component, the output jack is used. This circuit has a capacitor to block all but AC signals from entering the meter.

Measurement

The controls used when operating the meter are as follows:

- The FUNCTION switch selects the type of multimeter operation desired, such as DC +, DC -, or AC.
- The RANGE switch selects various voltage or resistance measurement ranges.
- The ohm's adjustment control, in the ohm position, is used to set the meter pointer to zero and compensates for internal resistor differences between the multipliers, as well as for battery strength.

VOLTAGE TESTERS

A "voltage tester" is found in almost every maintenance shop, Figure 2–28. It is often the first electrical testing device many technicians use. The voltage tester has many practical uses such as checking for blown fuses, open switches, open heating elements, energized or de-energized circuits, and indicating whether the circuit is AC or DC. These testers may give a general voltage reading, but they are not to be considered as accurate, and should never be used in the place of a quality voltmeter. These devices must be used carefully, and the manufacturers' guidelines must be followed. In the next section on high voltage, you will see that when used as live circuit indicators, they serve their purpose well.

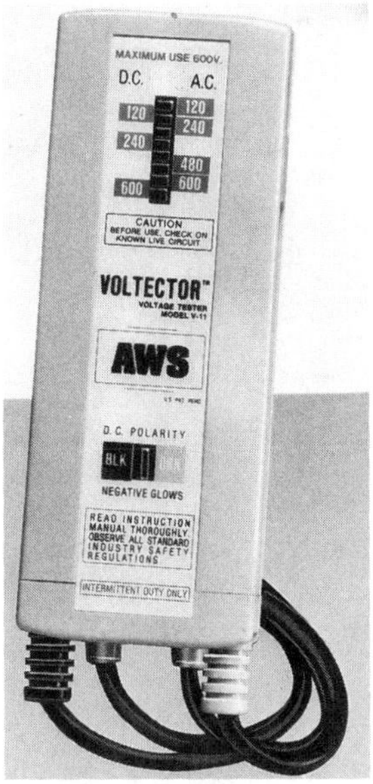

SPECIALIZED HIGH-VOLTAGE MEASURING INSTRUMENTS

The previous section dealt with analog testing equipment that will read up to 1,000 volts. In an industrial environment, voltages are used that sometimes exceed 13,800 volts, phase-to-phase (measured from energized lead to energized lead). You should at least be aware that a whole range of equipment is available to measure these voltages safely.

**FIGURE 2–28
Voltage tester.
(Courtesy of A. W.
Sperry Instruments,
Inc.)**

Some meters can tell the difference between a hot line and one that is de-energized using only one connection. Figure 2–29 is an example of a multi-range-voltage detector (MRVD) that can handle up to 40,000 volts. This MRVD unit uses the magnetic field surrounding the high-voltage line to tell if the circuit is live. Units using two connections are available that can read actual voltage values.

Most high-voltage units are based on a series resistor principle. A number of high value resistors are arranged in series and enclosed in an insulated material such as fiberglass. Although the insulating material has a high dielectric strength (high resistance to current flow), the occasional user, such as the industrial electrician, should also wear special high-voltage gloves which are rated to the voltage worked when taking any readings to ensure operator safety.

High-voltage workers use a sequence in taking measurements to ensure that failure of measuring equipment does not lead to a false sense of safety. Use this sequence whenever verifying that a line is de-energized.

**FIGURE 2–29
Multi-range voltage
detector. (Courtesy
of A.B. Chance
Company)**

1. Following OSHA guidelines, and using protective equipment on voltages over 50 volts, take an instrument reading from an energized circuit of the same voltage as the de-energized circuit. Make sure the measuring instrument reads the proper voltage.
2. Measure the circuit that is supposed to be de-energized. Make certain the reading is zero.
3. To be sure the measuring instrument did not burn out just as you took it off the first energized circuit (it has happened), go back and measure it again. If the meter indicates the same reading as in step number one, the second circuit is truly de-energized.

SUMMARY

All analog measuring elements and circuits use the same basic principle. They take current in one form or another and give readings for voltage, amperes, and ohms.

D'Arsonval movements and taut band measuring elements, a variation on the D'Arsonval movement, are most often used in general purpose test instruments. Their twin advantages of high accuracy and inexpensive construction make them the choice for most multimeters.

Because they operate on DC current, D'Arsonval movements must first rectify AC voltage. Readings, while given as rms, are really average responding.

Electrodynamometer, iron vane, and thermocouple elements are also used to measure voltage and current. While more expensive than other methods, they do have the advantage of measuring true AC rms voltage, and are not polarity sensitive.

Table 2-1 shows a performance comparison of the various types of metering elements.

Voltage is read by putting the meter in parallel with the load being measured.

Current requires that the meter be put in series with the measured load. However, devices are available that clamp around the load-carrying conductor and read current flow through induction, or using Hall effect devices.

High-voltage power circuits require special tools to read voltage values and assure that lines are de-energized. Hot sticks and high-voltage gloves are all designed to assure the safety of the test technician.

Table 2-1   Performance Comparison of Analog Metering Elements

Movement	Ruggedness	Response to DC	Response to AC
D'Arsonval	❷	❶	❷*
D'Arsonval taut band	❶	❶	❷*
D'Arsonval taut band w/ thermocouple	❶	❶	❶
Iron vane	❷	❸	❶
Electrodynamometer	❷	❸	❶

❶ Good   ❷ Fair   ❸ Poor
*only with Rectifiers

REVIEW QUESTIONS

1. It does not matter if a meter is reading voltage, current, or ohms. What deflects the D'Arsonval movement upscale?

2. Since AC voltage changes direction twice each cycle, why does the electrodynamometer needle not try to return to zero?

3. Since an ohmmeter needs voltage to read ohms, why does it matter if a measured circuit is energized or not?

4. What must AC voltage be changed to before a D'Arsonval movement can read the measured value?

5. When reading 100 amperes of current with an external shunt, why does it not destroy a meter element rated at 100 microamperes?

6. Why does the scale of a meter using a thermocouple have larger divisions the farther it goes up scale?

7. What is meter sensitivity measured in?

8. Why is accuracy for AC volts always less than that listed for DC volts?

9. An ohmmeter dial reads zero when the leads are shorted together. Of the current for which the meter element is rated, what portion of the current is going through the element?

10. Why should it be necessary to read the voltage of an energized circuit before and after checking a de-energized circuit?

Digital Hand-Held Test Equipment 3

INTRODUCTION

Until a few years ago, whenever digital test instruments were compared with analog meters, cost always made analog meters the choice. When both types of meters were judged on accuracy or ability to read test results, digital won easily. But few electrical maintenance facilities could justify the high cost of the digital meter. In many cases, they could buy two analog meters for the price of one digital unit.

But this scenario has changed. Now digital multimeters are available in all price ranges. Often models with many features are priced competitively with analog meters possessing similar features.

At one time, analog multimeters were the only ones that could test capacitance; now digital meters can do that too.

Need a meter that beeps for certain parameters? How about one that talks to you? All are now available in digital models.

OBJECTIVES

After studying this chapter, the student should be able to:

- *Describe a digital meter.*
- *Understand the steps all digital meters must go through to measure and display measured values.*
- *Explain the various ways digital meters change an analog value to a digital countable value.*
- *Understand the displays used to show test results.*
- *Know the difference between accuracy and sensitivity.*
- *Understand the uses for voltage detection devices.*

DIGITAL VERSUS ANALOG METERS

The instruments covered so far utilize analog measuring circuits and metering elements to measure voltage, current, and resistance. Analog merely means that the measuring device uses some physical forces to measure the relative changing value of voltage or current. For example, as voltage increases or decreases in intensity in a measured circuit, an analog meter responds by an interaction of magnetic forces. This interaction between measuring coils drives the pointer up or down scale, in response to the value of voltage measured.

Digital meters, on the other hand, do not use a scale, but display readings as numbers. While technicians take digital to mean a numerical display, digital meters are much more than that. It is also a system of taking physical elements, such as voltage, current, or ohms, and changing these to values that can be counted electronically.

You may recall that in the section on analog meters current was the means for measuring all values. Whether current, voltage, or ohms was being measured, current drove the pointer up scale.

With digital meters, a DC voltage is used to read out all values. It does not matter if voltage, current, or ohms are being measured; the method of measurement will be DC voltage.

Another difference is that analog meters usually use one, and sometimes two, batteries for the ohms measuring circuits. Digital meters use battery power both to measure and to power the display for all values.

THEORY OF OPERATION

In their simplest form, digital meters contain three major groupings of components. As shown in Figure 3–1, they are signal conditioners, analog/digital (A/D) converters, and display circuits.

Signal conditioners may be a set of selector switches and circuitry quite similar to analog meters. Various combinations of resistors are used to process incoming voltage and current signals, or to read out ohms. Then the DC signal can be passed on to the analog/digital converter.

Analog to digital converters use a number of schemes to measure voltage, current, and ohms. Many measure the charging time of a ca-

FIGURE 3–1
Digital meter processing sequence

pacitor or use other comparator methods. All rely on clock pulses counted over time, or a measured quantity such as internal voltage, to come up with a number, representing the input quantity measured.

Display circuits take the numerical results from the A/D converter and display them as values measured.

Signal Conditioners

Selector switches are used on some models to set the values to be measured. Volts, amps, or ohms have individual circuitry to preprocess the input element being measured. Ultimately, signal conditioners must either reduce the measured voltage or current down when reading high values, or amplify low values. From there the signal is handled and measured by the A/D converter.

Certain features such as polarity sensing are present at this stage. Unlike analog meters, there is no pointer to peg if polarity is incorrect. Instead, sensing circuits illuminate a "+" or "–" sign on the display to indicate if leads are measuring DC voltage as marked on the meter input terminals.

Another feature in the signal conditioning section on some models is autoranging. Output of an operational amplifier can be controlled by varying resistance between the output and the non-inverting minus input. The lower the resistor, the lower the output. As shown in Figure 3–2, one method of autoranging uses a number of resistors between the output and minus input of an operational amplifier. Field effect transistors (FET), themselves controlled by comparator circuits, add these resistors singly or in parallel, to automatically select the correct range to modify amplifier output. In this way, measured values will never exceed the capability of a given range.

ANALOG TO DIGITAL CONVERTER

A number of methods are used to convert analog voltage to a digital value. These include single slope, dual slope, staircase, voltage to frequency, and successive approximation. Each has certain advantages. Sometimes it comes down to which is the least expensive method to use to manufacture a certain model of meter.

SINGLE SLOPE CONVERTER

One of the simplest A/D converters is the single slope unit. As shown in Figure 3–3, the main components consist of a capacitor, crystal-controlled oscillator for clock pulses, precision voltage source, and voltage comparators.

FIGURE 3–2
Autoranging circuit

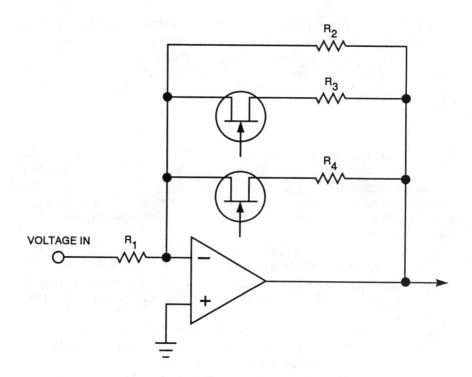

When a reset occurs at the end of a measured time period, voltage discharges past zero to a minus value. As the next sample of precision voltage is applied to the capacitor, voltage begins to rise. When voltage crosses from minus through the zero point, a comparator senses the crossing and starts a gating circuit. This allows clock pulses to enter the counting circuit. Once the precision voltage reaches the same value as the measured voltage, another comparator shuts off the counting circuit. Although the precision voltage continues to climb until the full preselected time period ends, no more clock pulses are accepted by the count circuit. Accumulated clock pulses will equal the measured voltage and are displayed. The capacitor is discharged and reset for the next sampling period.

Single slope converters are called non-integrating, in that they are not affected by noise or irregular sine waves, Figure 3–4. Measurements are taken in average voltage and read out as rms voltage.

DUAL SLOPE CONVERTER

A method used by the majority of better digital meters is the dual slope A/D converter, shown in Figure 3–5. Once again, a capacitor

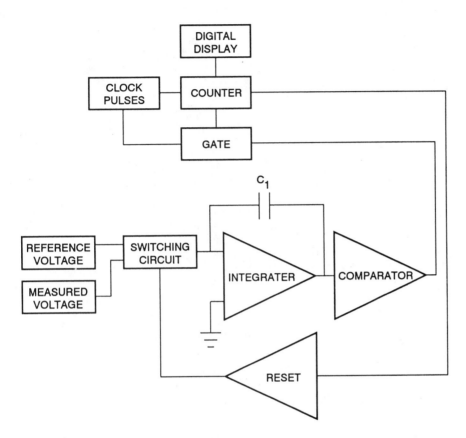

FIGURE 3–3
Slope methods of A/D conversion

is charged for a controlled period of time. However, the amount of charge will depend entirely on the magnitude of the voltage being measured. Higher voltage equals more charge. When the time period ends, a precision voltage of opposite polarity is applied to the capacitor, and it starts discharging. At the same instant, the counting circuit starts accepting clock pulses. When the capacitor is completely discharged and reaches zero, the counting circuit rejects all further clock pulses. Pulses accumulated equal the measured voltage and are shown on the display.

One of the advantages of the non-integrating dual slope converter is relative immunity to noise in the measured circuit. Voltage readings are given as rms, but are actually average voltage readings.

The disadvantage is that the dual slope converter is one of the slowest digital methods of measurement.

FIGURE 3–4
Single slope A/D
converter

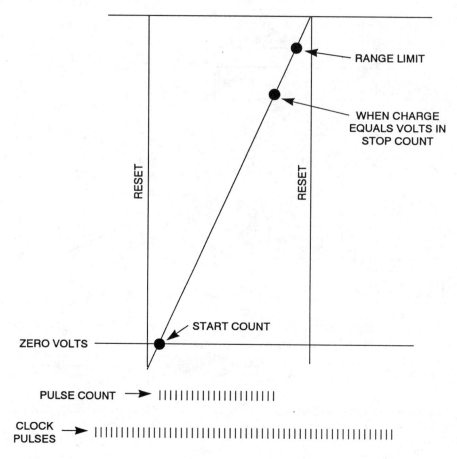

RANGE LIMIT

WHEN CHARGE
EQUALS VOLTS IN
STOP COUNT

RESET

RESET

START COUNT

ZERO VOLTS

PULSE COUNT ⟶ |||||||||||||||||||||||||

CLOCK
PULSES ⟶ |||||||||||||||||||||||||||||||||||||||||||||||||||||

FIGURE 3–5
Dual slope A/D
converter

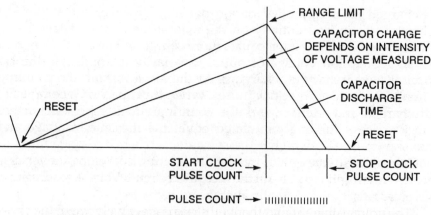

RANGE LIMIT

CAPACITOR CHARGE
DEPENDS ON INTENSITY
OF VOLTAGE MEASURED

CAPACITOR
DISCHARGE
TIME

RESET

RESET

START CLOCK
PULSE COUNT

STOP CLOCK
PULSE COUNT

PULSE COUNT ⟶ ||||||||||||||||||

CLOCK PULSES ⟶ ||||||||||||||||||||||||||||||||||||||||||||||||||||||||

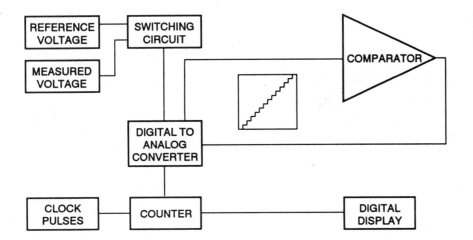

FIGURE 3–6
Staircase converter

STAIRCASE RAMP CONVERTER

In the staircase ramp converter, an internal precision voltage is generated in increasing increments, or steps over time, and compared to the
measured voltage. Figure 3–6 shows how the measurement takes place.
When the measured voltage and stepped voltage are both at zero, the
count circuit is gated. Stepped voltage increases at measured amounts
until it matches measured voltage. At that point, the count gate is
closed, and pulses accumulated equal voltage input. Because the count
represents a voltage value, the converter is actually a digital to analog
unit (D/A).

VOLTAGE TO FREQUENCY CONVERTER

Voltage to frequency converters also use a charging capacitor to
measure voltage input, as shown in Figure 3–7. As the cycle starts,
measured voltage is applied and charges a capacitor. Once capacitor
voltage equals reference voltage, a comparator triggers discharge of
the capacitor to zero voltage, generates a pulse, and starts the next
charging cycle. The higher the measured voltage, the more times the
capacitor will charge and discharge, which also means more pulses
over time. Total pulses equal the measured voltage, as shown in
Figure 3–8.

The voltage to frequency converter also is relatively immune to noise
and reads average voltage displayed as rms voltage.

FIGURE 3–7
Voltage to frequency
converter

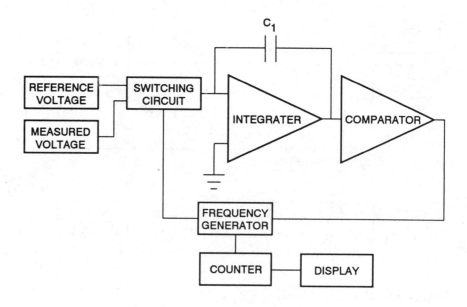

SUCCESSIVE APPROXIMATION CONVERTER

This system of conversion uses a binary-coded decimal (BCD) sampling technique to read out measured values, as shown in Figure 3–9. Digital counting circuits are either off or on. Off is zero (0), and on is one (1). In order to count in a binary system, a series of positions increasing by the power of two are arranged from right to left. Power of two

FIGURE 3–8
Voltage to frequency
A/D converter

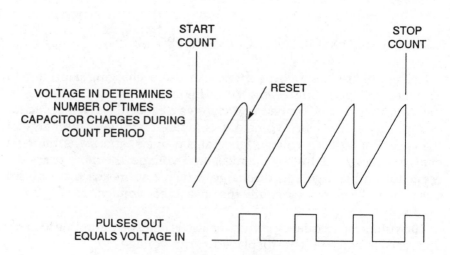

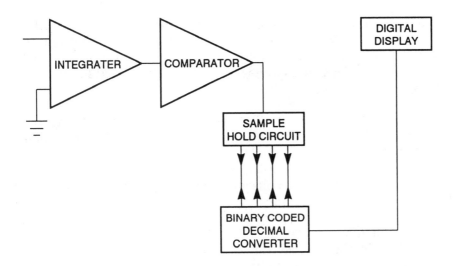

FIGURE 3–9
Successive
approximation circuit

merely means that each position to the left is doubled in value. Therefore, the following values exist between positions:

BCD position		0	0	0	0
Numerical value	8	4	2	1	

The number one (1) written in BCD is 0001; three (3) would be 0011;
and thirteen (13) would be 1101, and so on.

In the successive approximation method, a sample of measured voltage is compared with an internally generated, increasingly smaller,
reference voltage. Digital/analog converters (D/A), four for units, four
for tenths (0.1), and four for hundreds (0.01), compare the sample
against the precision voltage at each binary position, as in Figure 3–10.
Notice that the groupings of binary code are in decreasing order from
left to right.

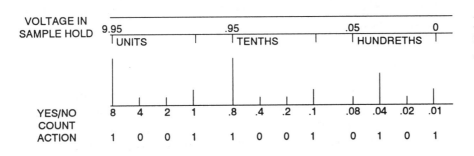

FIGURE 3–10
Successive
approximation

An example of how the successive approximation converter works is best shown by following the steps taken to read out a voltage, in this case 9.95. To make it easier to follow the sequence of events, MV will be used to indicate measured voltage, and RV will be used for reference voltage.

1. The first step is to put the 9.95 volts in a holding circuit. A comparison between MV and RV shows that MV is more than 8, so that position is turned on. Eight volts are subtracted from both MV and RV.
2. Remainder equals 1.95 volts
3. Remaining MV of 1.95 volts is compared with 4 volts RV and is not accepted, so this position remains off. Since MV is still not higher than 2 volts RV, that position also remains off. At the 1 volt RV position, MV is higher than RV, so the 1 RV position is turned on, and 1 volt is subtracted from both MV and RV.
4. Remainder equals 0.95 volts.
5. MV is compared with 0.8 volts RV, is accepted, the position is turned on, and 0.8 volt is subtracted from both MV and RV.
6. Remainder equals 0.15 volts.
7. MV is compared and rejected at the 0.4 volt and 0.2 volt RV positions, and they remain off. At 0.1 volt RV, MV is higher, so that position is turned on, and 0.1 volt is subtracted from both MV and RV.
8. Remainder equals 0.05 volts.
9. At 0.08 volts RV, the position remains off because RV volts are higher than MV. However, 0.04 volt RV position is turned on, as MV is higher than RV, and 0.01 is subtracted from both voltages.
10. Remainder equals 0.01 volts.
11. RV position 0.02 volts is rejected and remains off. Position 0.01 is turned on, because RV voltage and MV are equal; and there is no remainder.
12. BCD value of 1001 1001 0101 is counted and displayed as 9.95 volts.

Because there are only twelve steps in the above calculations, the successive approximation conversion method is one of the fastest in operation. It is more costly to build and is usually used on better meters. Because it is sensitive to noise, filtering is necessary to condition the incoming signals.

THERMOCOUPLE CONVERTERS

One method that is used to read out true rms readings in digital meters is by using a thermocouple. Remember that the principle covered in Chapter 2 showed that if two dissimilar metals are welded together and heated by a device that sensed current, a voltage is given off. Because

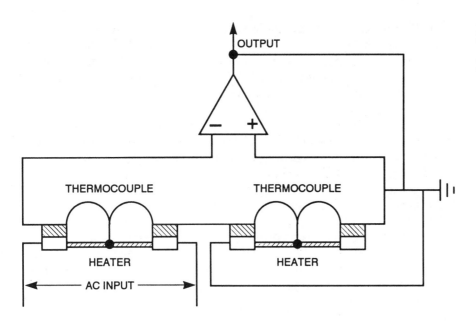

**FIGURE 3–11
Thermocouple
processor**

heat was involved, the unit operated on a current squared (I^2) principle. As a result the travel up scale was not linear.

In order to design a circuit for digital meters some variation on the principle must be used. Figure 3–11 shows two thermocouples connected together. Using a feedback principle through the operational amplifier not only makes the output linear, but prevents surrounding temperatures from affecting the unit as well.

DIGITAL READOUTS

The advantages of digital readout indicators have led to their widespread adoption. The reading errors associated with magnetic meter movements are virtually eliminated with digital indicator readouts, since the measurements appear as a number on a numerical (digital) display.

The most common types of digital readouts are gaseous discharge tubes, matrix displays, dot or bar segments, light-emitting diodes (LED), and liquid crystal display (LCD).

Gaseous Discharge Tubes

Gaseous discharge tubes were once widely used electronic alpha/numeric readouts in which the mode of operation is similar to the neon lamp. Standard numerical display tubes contain a common anode and a stack of ten independent cathodes shaped like numbers. When a large negative voltage is applied to a selected cathode, the gas around it glows from being ionized and emits an orange light. If the inmost cathode is lighted, for instance, it is slightly recessed, and the observer looks through the remaining unlighted cathodes above it.

Alpha/numeric readout tubes are similar to numerical tubes except that all the letters of the alphabet and numbers 0 to 9 can be generated by a single-plane, fourteen-segment, bar matrix. The ends of the segments are closely spaced, and sharp character resolution is obtained without the noticeable gaps found in other types of segmented displays.

Gaseous discharge tubes require a high-voltage DC power supply to ionize the neon gas. The positive side of the power supply is connected (in series) through a current-limiting resistor to the common anode. An integrated circuit decoder with memory accepts binary coded decimal inputs, connects the appropriate cathode to the negative side of the power supply, and keeps the cathodes lighted.

Matrix Displays

Matrix displays are digital readouts forming characters by selectively lighting dots or bar segments from within a matrix, Figure 3–12. Standard dot matrices are 5 × 7, 6 × 8 and 7 × 9, and can display all letters and numbers. The 7 × 9 matrix is preferred, since it has best character resolution. Standard bar matrices use 7-, 12-, 14-, and 16-bar segments. Only the last-named can form all alphabetic and numeric characters without ambiguity. Matrix displays provide only fair to good character

FIGURE 3–12
Digital readouts

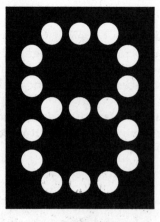

5 x 7 DOT MATRIX   7 BAR SEGMENT MATRIX

legibility. For example, the numbers 0, 1, and 2 can be misinterpreted as the letters O, I, and Z, respectively.

Dots or Bar Segments

Dots or bar segments can be lighted by incandescent, neon, fluorescent, or solid-state lamps. Incandescent lamps light dot matrices through fiber optic light guides and form high contrast characters. They can also be placed directly behind small acrylic plates, acting as bar segments. Individual neon or fluorescent lamps can be used as the actual bar segments. Solid-state lamps are usually arranged behind a filter as separate dots or bars.

Light-Emitting Diode

The solid-state lamp is commonly called a light-emitting diode (LED), and it is a valuable by-product of semiconductor technology. It is basically a P–N junction diode mounted in a hermetically sealed case, with a lens opening at one end. Light is produced at the junction of the P and N materials by two steps. A low voltage DC source increases the energy level of electrons on one side of the junction. To maintain equilibrium, the electrons must return to their original state; they cross the junction and give off their excess energy as light and heat. The light output of the popular gallium arsenide phosphide LED has a very narrow bandwidth centered in the red band. These lamps are very efficient and have exceptionally long life, but are small and have low light output.

Liquid Crystal Displays

Liquid crystal displays (LCDs) also utilize a seven segment display; ambient or surrounding light, usually reflected from a silvery background or absorbed by an activated dark segment, makes the characters visible.

A completely metalized rear glass plate and a front glass plate (with only seven segments of each character metalized) form a sandwich, trapping the liquid crystal between them. Two low-voltage AC signals are sent to the metalized portions of the rear and front plates. Any segments on the front plate whose signals are in-phase with the rear plate show up as silver and are not visible. Those segments out-of-phase become visible as dark areas.

DIGITAL MULTIMETERS

In the section on analog meters, a variety of instruments were covered. These ranged from panel meters to various hand-held instruments. When digital meters are discussed, they too range in applications.

However, one of the most popular instruments used in the electrical maintenance field is the digital multimeter. A simplified circuit for this meter is shown in Figure 3–13.

Figure 3–14 is typical of the models available today. Primarily these units will measure DC volts and amps, ohms, and AC volts and amps. Other features such as autoranging and polarity sensing extend the usefulness of the meter. Some models include a bar graph that can simulate the dial pointer of an analog meter. This feature comes in handy when a widely varying signal being measured makes it difficult to read the digital display.

DC Voltmeter

**FIGURE 3–13
Digital multimeter
circuit**

This might be a good time to review the sequence used by high-voltage workers in taking measurements. It ensures that failure of measuring equipment does not lead to a false sense of safety. Use it in all cases where checking to verify that a line is de-energized.

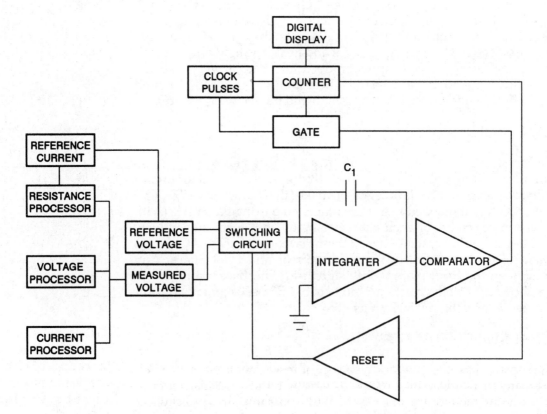

1. Following OSHA guidelines, and using protective equipment on voltages over 50 volts, take an instrument reading from an energized line of the same voltage, other than the one you will measure. Make sure the measuring instrument reads the proper value.
2. Measure the line that is supposed to be de-energized. Notice if reading is zero.
3. To make certain the measuring instrument did not burn out just as you took it off the first energized circuit (it has happened), go back and measure it again. If the meter indicates the same reading as in step number one, the second circuit is truly de-energized.

> WARNING Treat every circuit as a live circuit.

Digital voltmeters usually use some variation of a voltage divider circuit, as shown in Figure 3–15. Just as resistors were used in analog measuring circuits to limit current through the meter element, the voltage divider circuit limits voltage to the analog to digital converters (or digital to analog circuit).

Because DC is a steady state voltage, it has a continuous direction of polarity. Many digital meters have the ability to read DC voltage, regardless of lead connection. Readings on the display will show a plus (+) or minus (–) sign, to let the user know if the lead connection is as marked on the instrument panel.

Although most meters will read out to 1,000 volts DC, keep in mind the warning about voltage levels given in Chapter 1. Safe operation of meters will be covered in detail in Chapter 4.

In cases where DC voltage exceeds the range of the meter scales, high-voltage probes, such as shown in Figure 3–16, are available. These probes are constructed of a string of dropping resistors, enclosed in a shockproof handle. They are not intended for use on AC voltage power circuits. Their use is primarily restricted to television picture tube or automobile spark plug voltage measurements.

DC AMMETERS

Ammeters are used by inserting the instrument in series with the load being measured. Internally, measurements are taken by passing current through various range resistors, as shown in Figure 3–17. Voltage drop across the resistor is read by the analog to digital converter, and displayed as a current value.

Manufacturers offer an accessory clamp-on device that will read out AC or DC current up to 200 amperes, Figure 3–18. Called a Hall effect AC/DC current probe, this battery-operated device does not work on a transformer principle, which would be useless for DC current. Instead, semiconductor material is embedded within the jaws. A DC voltage,

FIGURE 3–14
Digital multimeter. (Reproduced with permission of John Fluke Mfg. Co., Inc.)

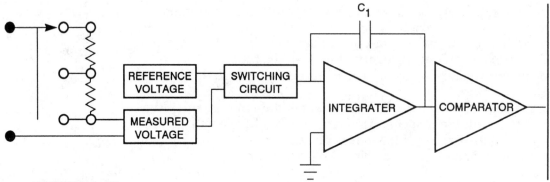

FIGURE 3–15
Voltmeter circuit

FIGURE 3–16
High-voltage probe.
(Reproduced with
permission of John
Fluke Mfg. Co., Inc.)

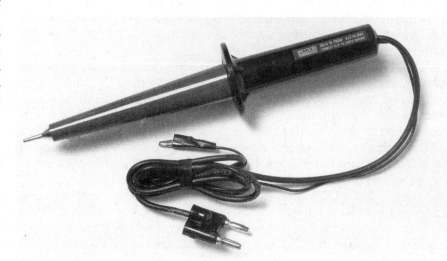

FIGURE 3–17
Ammeter circuit

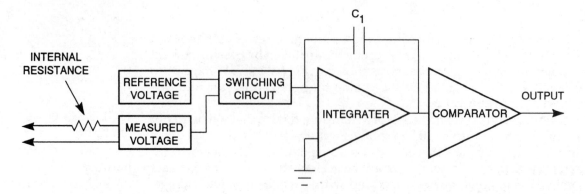

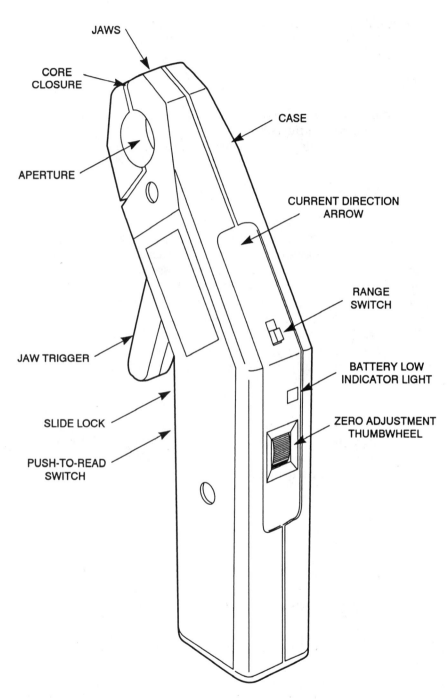

FIGURE 3–18
DC/AC current probe. (Reproduced with permission of John Fluke Mfg. Co., Inc.)

proportional to the current measured, is passed on to the meter. The unit is switch selectable and will develop two-volts output at the 20- or 200-ampere position.

OHMMETERS

Digital ohmmeters use a constant current passing through the unknown resistance and measure the voltage drop across it, as shown in Figure 3–19. Digital display and measuring techniques allow readings down to 0.1 ohm.

AC VOLTMETER

Whether it is an analog or digital voltmeter, AC must be changed to DC voltage before measurement can take place. The usual method is to pass voltage through a full wave rectifier, before passing the DC voltage on to the analog to digital converter. This extra step of rectifying AC voltage to DC, before processing by the analog to digital converter, means AC voltage readings will never be as accurate as DC voltage readings.

AC voltage ranges can be up to 1,000 volts or higher. Again, remember the warning about handling voltages above 50 volts without protective gear, such as rubber gloves. The best method for taking voltage readings is to turn off apparatus, connect meter, then turn apparatus on.

AC AMPERES

Uses circuitry that is similar to DC ammeter. Accessory clamp-on probes are available that allow AC measurement up to 400 amperes.

Clamp-on ammeters are now available that can measure peak, average, or rms values of current at the turn of a switch. Shown in Figure

FIGURE 3–19
Ohmmeter circuit

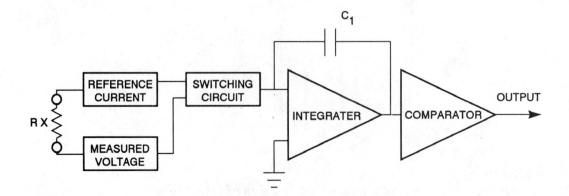

FIGURE 3–20
Peak-RMS-average
reading clamp-on
ammeter. (Courtesy
of Amprobe
Instrument®)

ACD- 2000

ACD- 2001

3–20, they serve the needs of a variety of personnel, from the electrical engineer to the maintenance technician.

DIGITAL METER SPECIFICATIONS

Peak voltage, rms, average voltage: you know the difference now. What do all the advertisements for digital meters mean? One meter will spell out rms display, average reading meter. Another will list rms display, peak reading. Still another will advertise a true rms reading meter.

You will have to look over the specifications very carefully if you truly need an rms responding meter. Usually they will say "true rms reading." This is the only type that will give accurate voltage readings on a distorted sine wave, such as might be found in circuits loaded down with computers. It will not be inexpensive.

Do you really need a true rms reading meter? Not if you use them to tell the difference between a 208- and 240-volt polyphase circuit, or if you only need to know the difference between a live and de-energized circuit.

Range

Most of the newest digital meters use a seven-segment display to read out numbers. All numbers from zero (0) to nine (9) can be formed with various combinations. On a three-digit display, any number from 1 to 999 can be displayed and easily read. On a four-digit meter, 9999 can be displayed. Suppose a two-segment display were added to the far left in both meters? The only digit it could display would be 1. But how powerful that ability is. Now the 3 1/2-digit meter can display 1,999 or the 4 1/2-digit meter could display 19,999. With the addition of that half-segment, and appropriate circuitry, both meters have nearly doubled the capacity for each meter range.

Accuracy

Accuracy is the ability of the meter to sense what the measured value actually is. It is usually given as a percent of the reading, plus or minus some value of the lowest significant digit (LSD). For example, one popular model lists accuracy for its DC current ranges as 0.8 percent of reading, plus or minus one LSD.

Suppose a five-ampere load is read on the ten-ampere range. This means that the meter circuitry could measure a five-ampere load, and calculate the value 4.96 to 5.04. Because of rounding off the least significant figure, the display could read 4.95 or 4.97, up to 5.03 or 5.05.

Resolution

Resolution is the difference in values that a meter can display. In a 3 1/2-digit meter, one part in 1,000 can be displayed, so 1/1000 = .001. Multiplying .001 times full scale of ten amperes, for example, equals a resolution of 0.01 amperes.

LIVE CIRCUIT TRACERS

A number of digital voltage tracers on the market can aid in identifying live circuits, such as the unit shown in Figure 3–21. When held to an energized conductor or wall outlet, it gives off a beeping signal. Not only are these devices useful for safety, but they can trace breaks in insulated wires and heating elements as well. They work on the principle of a small radio receiver that is sensitive to the 50 through 60 cycle electrical signal, as shown in Figure 3–22.

Units are available that can read from 30 to 122,000 volts. Although insulated hot stick handles are available for safe reading, protective rubber gloves rated for those voltages should also be worn by the industrial electrician.

Care must be taken in measuring three-phase circuits where all conductors are close together. Signals may be misleading as to which is the live and which is the de-energized conductor.

FIGURE 3–21
Voltage indicator.
(Courtesy of TIF
Instruments, Inc.)

SUMMARY

The price of digital meters is now more competitive with that of analog instruments. Digital meters must first change analog values, such as volts, ohms, and amperes, to digital values that can be counted and displayed. A variety of methods are used for this conversion, but usually involve a measurement of voltage. This is exactly the opposite of analog meters, which usually measure some current value.

Circuit measurement is quite similar to methods used by analog instruments. Voltage is always measured with the meter in parallel. Current is measured with the instrument in series. Ohms can only be read on de-energized circuits.

Some models of digital meters offer autoranging and polarity sensing circuits, to simplify use and prevent overloading.

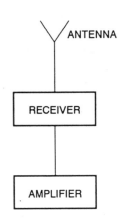

FIGURE 3–22
Live circuit indicator

REVIEW QUESTIONS

1. If the analog D'Arsonval movement uses DC amperes to measure all values, what does the digital meter circuitry use?

2. What are the three steps all digital meters go through to measure and finally display volts, amps, and ohms?

3. How is a device as simple as a capacitor able to make such complex measurements in a digital meter?

4. How many segments are used in LCD digital displays to make numbers from 0–9?

5. What is meant by 3 1/2 or 4 1/2 digits?

6. Why do digital meters impose less of a load on measured circuits than analog meters do?

7. When inexpensive digital meters give readings in rms, what are they actually measuring?

8. How do digital meters use current to read resistance?

9. What is meant by accuracy?

10. What is resolution in a digital meter?

Care and Application of Meter Test Equipment

4

INTRODUCTION

Safety is of primary importance at any time, but especially for those who work with electrical circuits. OSHA has developed guidelines for voltage levels that can be handled without protective gear. Surprisingly, that voltage is below 50 volts.

Carelessness is one of the major causes of test equipment failure. Equipment downtime can be minimized simply by ensuring that the proper meter is used for the measurement needed.

In this chapter electrical safety, along with calibration systems, meter accuracy, and meter applications, will be covered.

Ammeters, voltmeters, and ohmmeters are important test instruments. An ammeter is used to measure current and should be connected in series with the device or circuit to be measured. A voltmeter is used to measure the potential difference between two points and should be connected in parallel with the device or circuit to be measured. The ohmmeter is used to measure resistance of de-energized circuits in values up to megohms.

Meters have come a long way in the past few years. Digital meters have a variety of new features, from a bar graph to models that give the meter readings in a human-sounding voice. Nevertheless, analog models have not left the scene. Most manufacturer equipment catalogs show units that have been favorites for years.

OBJECTIVES

After studying this chapter, the student should be able to:

- *Understand the OSHA regulations that affect electrical safety.*
- *State the major cause of test equipment failure.*

- *Identify a meter in calibration.*
- *Define meter accuracy.*
- *State how to zero a meter.*
- *Define parallax error.*
- *Identify the proper application and connection of test equipment.*
- *List two methods used in extending meter ranges.*

SAFETY

The purpose of this chapter is to learn how to treat and use portable meters for taking measurements. Before getting into applications, which involve working with potentially dangerous circuits, it might be best to go over some of the safety regulations that affect the electrical industry. Chapter 1 gave some idea of what is expected of workers and those in charge of them. This section will also deal with some safety equipment and its care.

As mentioned earlier, most electrical workers have total respect for voltages of 240, 480, or higher. There is no doubt in their minds that these voltages can be deadly if bare skin makes contact with them. However, it is a matter of record that most deaths in the electrical industry are caused by 120-volt circuits. It is also true that most accidents occur to utility workers, who are not always trained to work with electrical circuits. This paragraph is not meant to frighten you, but to make you aware of the problems.

That is why the Department of Labor, under its Occupational Safety and Health Administration (OSHA), has issued rules for personnel exposed to possible contact with voltages of 50 volts or higher. In Document 29 CFR Part 1910, OSHA says that, if employees are in a position to make contact with uninsulated live circuits of 50 or more volts, they must wear protective equipment, such as rubber gloves, and in certain cases rubber sleeves as well. They must also receive instructions in how to safely work in that area.

Exposed is the key word in the statement. Does a housewife plugging in an appliance or a child turning on a light switch need rubber gloves? After all, in most cases the voltage is at least 120 volts, and people have been shocked by touching the metal prongs when inserting a plug into a receptacle.

The answer is no. These conditions are normally not considered as being exposed. If a technician's duties include working on the back of panels where fingers are a few inches away from bare, energized receptacle terminals, that is being exposed. In the technician's case, rubber gloves would be called for.

Rubber gloves come in a number of voltage ratings:

Class 0. For handling circuits up to 1,000 volts. Shown in Figure 4–1A. These gloves are designed for meter personnel use and other low-voltage applications. (Note: Low voltage in the electrical industry is any voltage up to 1,000 volts AC.)
Class 1. For voltages up to 7,500 volts AC, as shown in Figure 4–1B.
Class 2. For voltages up to 17,000 volts AC.
Class 3. For voltages up to 26,500 volts AC.
Class 4. For voltages up to 36,000 volts AC.

**FIGURE 4–1A
Class "0" gloves.
(Courtesy of North
Hand Protection)**

Rubber gloves are relatively soft, so they are never worn by themselves. Instead, leather protectors, as shown in Figure 4–2, are usually used. Also, special canvas glove bags hold the gloves when not actually in use to avoid accidental damage.

There are industry standards for testing rubber gloves. For example, Class 0, although rated at 1,000 volts, are tested at 5,000 volts AC. Usually outside electrical laboratories are used to test gloves on a scheduled basis. Between scheduled tests, it is up to the user to make certain that the gloves are electrically safe. Figure 4–3 shows visual and leak tests that can be performed before each use.

HOT STICKS

Fiberglass handles are often used when tests are to be made on energized circuits. Rules governing distance from energized circuits for *qualified personnel* can get a bit involved.

Note the key words are *qualified personnel*. If the technician is trained in electrical work, the guide for distances to approach overhead lines are as follows (as quoted in Table S-5, OSHA publication 29 CFR Part 1910 of the *Federal Register*, dated August 6, 1990):

Phase-to-Phase Voltage	*Minimum Approach Distance*
300 V or less	Avoid contact
Over 300 V; not over 750 V	1 ft. 0 in.
Over 750 V; not over 2 kV	1 ft. 6 in.
Over 2 K; not over 15 kV	2 ft. 0 in.
Over 15 k; not over 37 kV	3 ft. 0 in.

Hot sticks are available for small hand-held instruments as shown in Figure 4–4 (page 82). One manufacturer offers two and four foot collapsible lengths. Electrical utilities have rules governing when rubber protective equipment must be worn while using hot sticks. However, for the industrial electrician, a good rule of thumb is to always wear protective equipment of a rating to match the circuit voltage.

**FIGURE 4–1B
High-voltage rubber
gloves. (Courtesy of
North Hand
Protection)**

SAFETY AND METER LEADS

Many multimeters have voltage ranges of 1,000 volts. Insulation covering the lead wires is normally rated at 600 volts, so it is not a good idea to put complete trust in this insulation when working with high voltages.

FIGURE 4–2
Gloves and
protectors. (Courtesy
of North Hand
Protection)

Fortunately, many instrument leads have probe tips embedded in plastic handles that will allow the user to carefully take measurements without having to resort to rubber gloves every time a reading over 50 volts is attempted.

If there is any doubt about safety, turn the circuit off. Attach the test leads to the circuit to be measured; then turn the circuit back on. Repeat the process to remove the leads.

Again, reviewing the sequence used by high-voltage workers in reading voltages or checking that a circuit is de-energized:

> **WARNING**
> *Treat every circuit as a live circuit.*

1. Following OSHA guidelines and using protective equipment on voltages over 50 volts, take an instrument reading from an energized circuit of the same voltage as the de-energized circuits. Make sure the measuring instrument reads the proper voltage.
2. Measure the line that is supposed to be de-energized. Make certain the reading is zero.
3. To be sure the measuring instrument did not burn out just as you took it off the first energized circuit (it has happened), go back and measure it again. If the meter indicates the same reading as in step number one, the second circuit is truly de-energized.

Reversing Gloves for Inspection

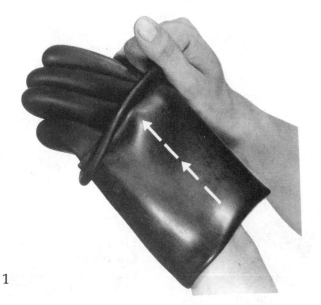

1

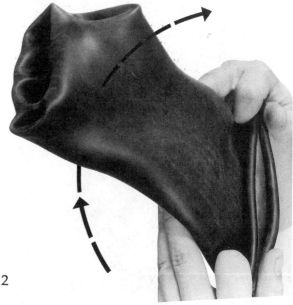

2

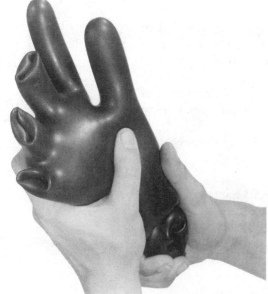

3

Reversing gloves for inspection can be accomplished by grasping the cuff and pulling it over the fingers as shown in illustration 1.

Then, holding the glove downward, grasp the cuff as in illustration 2 and twirl it upward toward your body to close cuff.

Next, squeeze the rolled cuff into a U shape to trap the air inside the glove and pop out the fingers by squeezing the inflated glove (3).

AFTER INSPECTION, TURN THE GLOVE RIGHT SIDE OUT!

To air test gloves for pinholes and other damage, follow these procedures: Hold the glove downward and grasp the cuff as shown in illustration 4.

Twirl the glove upward toward your body to trap the air inside the glove as shown in illustration 5.

Then squeeze the rolled cuff tightly into a U shape with the right hand to keep trapped air inside. Squeeze with the other hand and look for damage exposed by inflation as shown in illustration 6.

Then hold the inflated glove close to your face and ear, squeezing the glove, to feel and listen for air escaping from holes as shown in illustration 7.

IMPORTANT PRACTICES TO FOLLOW:

INSPECT FREQUENTLY FOR GLOVE DAMAGE. RINSE GLOVES DAILY WITH CLEAR WATER INSIDE AND OUT AND THOROUGHLY DRY. WEAR LEATHER PROTECTORS. REPLACE OLD, WORN OR DAMAGED PROTECTORS.

GIVE NORTH SLEEVES THE SAME PROPER CARE YOU GIVE NORTH GLOVES.

FIGURE 4–3
Rubber glove self-test (Courtesy of North Hand Protection)

Inspecting Gloves for Small Leaks

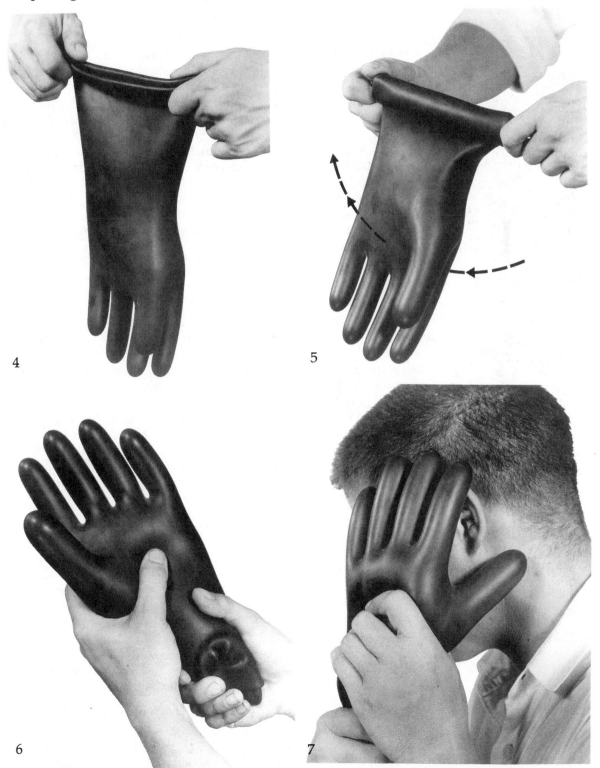

4

5

6

7

TREATMENT OF METER TEST EQUIPMENT

Electrical maintenance shops are provided with a variety of test equipment used in testing and troubleshooting the equipment they support. Seldom are there many spare test sets, so when test equipment is damaged, shop maintenance slows down. Therefore, to avoid problems, use the test equipment for the specific purpose for which it was designed.

One of the chief causes of test equipment failure is careless operating procedures. Mistakes such as trying to measure a 100-volt circuit on the 10-volt scale, or reading ohms in an energized circuit, often lead to instrument downtime. Technicians often place test sets near the edge of a bench where they can easily be pulled off by the test leads. To avoid damage, be aware of potential hazards, and read the instruction manual for proper operation.

WEATHER EXPOSURE EFFECTS

Some test equipment requires special handling. However, several precautions apply to test equipment in general. Rough handling, moisture, and dust all affect the useful life of these devices. Bumping or dropping a test instrument, for example, may destroy the calibration of a meter or short circuit the elements within the instrument.

To reduce the danger of corrosion to untreated parts, always store test equipment in a dry place when not in use. Excessive dust and grime inside a test set affects the accuracy. Be sure that all screws holding the case of the test equipment in place are tightened securely. As an added precaution, dust covers should be placed on test equipment not in use. Test equipment is delicate and must be treated properly; the major cause of test equipment failure is carelessness.

**FIGURE 4–4
Collapsible hot stick
(Courtesy of TIF
Instruments, Inc.)**

CALIBRATION SYSTEM

Calibration is the process of comparing a measuring device with a standard. A standard is the measurement that is widely recognized as the most accurate measurement of a particular unit. Whether the standard is a foot rule, clock, scale, voltmeter, or any of numerous measuring devices, a standard exists for each value to be measured.

The National Institute of Standards and Technology (formerly called the National Bureau of Standards) provides standards and services for the United States. Their calibration program establishes a link between standards maintained by them and test equipment used by test personnel in the field.

When any piece of equipment is calibrated, a sticker is placed in a conspicuous place on the instrument. This sticker states the date of calibration and the date the instrument is due for calibration again. To prevent tampering with adjustments, a seal is also placed on the instrument. If the calibration sticker or seal has been tampered with, the instrument should be returned to the calibration lab to verify its accuracy.

Every state in the United States has one or more electrical laboratories that have test equipment traceable to the NIST, should the need arise to have equipment certified.

Adjusting Zero

Unless an analog meter starts from a true zero position, no measurement will be accurate. One suggested method for adjusting zero is as follows:

   With the volt-ohm milliammeter in an operating position, check that the pointer indicates zero at the left end of the scale when there is no input.

   If the pointer is off zero, adjust the screw located in the case below the center of the dial. Use a small screwdriver to turn the screw slowly, clockwise or counterclockwise, until the pointer is exactly over the zero mark at the left end of the scale.

   With the indicating pointer set on the zero mark, reverse the direction of rotation of the zero adjuster.

   Rotate the zero adjuster (being careful not to disturb the position of the indicating pointer) a sufficient amount to introduce mechanical freedom, or "play."

This procedure will avoid disturbances to the zero setting by subsequent changes in temperature, humidity, vibration, and other environmental conditions.

READING METERS

In reading analog meters, understanding a few simple procedures will ensure that the values read are as accurate as possible.

Parallax

When reading analog meters, the prospect always exists that two technicians will read different values from the same scale plate. The problem becomes worse if either reads the meter from the side. If not directly over the pointer, the eyes see the side of the pointer, and readings will be either higher or lower than the actual reading. This error is called parallax.

Two common methods are used to eliminate parallax error. One is to twist the tip of the pointer to form a very thin line, or knife edge. The meter is read so only the thin top edge of the pointer is viewed. If the side of the knife edge is visible, the eyes are not exactly over the pointer.

Another method is to use a mirrored scale. When the pointer covers its own reflection in the mirror, it means the eyes are exactly over the pointer. If viewed from the side, the pointer and its reflection will both be visible. Some meters incorporate both a knife edge pointer and a mirror scale to help eliminate parallax error.

Meter Accuracy

Meter accuracy may be stated as either percent of reading or percent of full scale. For example, a meter that has an accuracy of ± 1 percent of reading will show a value that is within 1 percent of the correct value. This means that if the correct value is 100 units, the meter indications may be within the range of 99 to 101 units.

Voltage or current accuracy of instruments also can be expressed as percent of full scale. This should not be confused with accuracy of reading (indication). For example, ±2 percent of full scale on a ten-volt range allows an error of ±0.20 volts at any point on the dial. This means that at full scale the accuracy of the reading would be ±2 percent. At half scale that same .02 error would equal ±4 percent. Therefore, it is best to select a range where readings indicate as near as possible to full scale.

Remember, accuracy of digital meters is stated as a percentage of full scale, plus or minus one digit, as explained in Chapter 2.

Application

In this section, some rules of operation and typical uses for volt-ohm-ammeters will be suggested. Although drawings may show analog meters being used, remember that there are digital meters available as well.

In order to alert the user to possible dangers when taking readings, the word WARNING will appear at the beginning of each instrument description. This method is being used by most equipment manufacturers. It is just another reminder about live circuits and their potential for personal injury.

In a facility such as a steel mill, DC is used for cranes and other motors. AC and DC circuits are usually separated and well marked. If there is any doubt, however, a voltage indicator, as described in Chapter 2, will show whether the circuit is AC or DC.

DC AMMETERS

DC in-line ammeters are used by observing polarity and putting the meter in series with the circuit being measured, as shown in Figure 4–5. If the DC ammeter is accidentally put in parallel across the load, a dead short circuit will result. Either the meter, its measuring circuit, or both can be destroyed.

Make sure the range selected is of sufficient size to handle the load. If in doubt, start with the highest range and select lower ranges until readings are in the upper quarter of the meter scale in analog meters. The same holds true for digital meters, except that the half digit gives the user more flexibility.

Loading down the circuits being measured (the meter circuit adds to the measured load) is usually not a serious problem with digital meters or better analog meters, but be aware that it can happen.

• If the current to be measured is beyond the highest range of the meter, two options are open. A shunt, as shown in Figure 4–6, can be used with the meter. Or AC/DC current probes are available that extend the range of meters to hundreds of amperes.

> *WARNING*
> *Treat every circuit*
> *as a live circuit.*

DC VOLTMETERS

DC voltmeters are used in parallel and observe polarity, as shown in Figure 4–7. If accidentally put in series with a load, the instrument will not be able to carry the current through the measuring circuit, and damage to the meter will result.

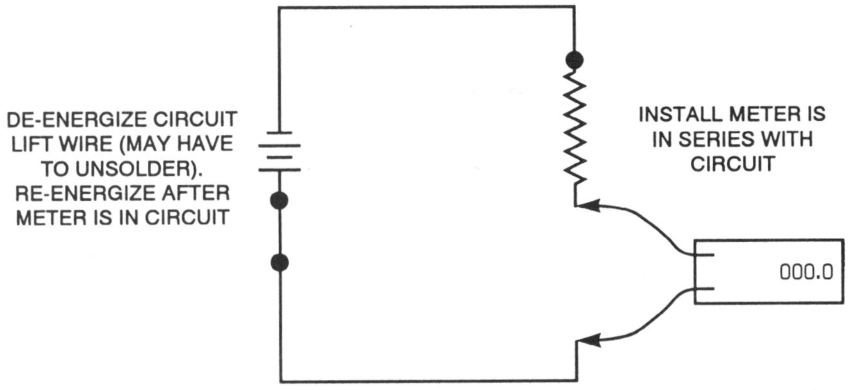

FIGURE 4–5
DC ammeter

DE-ENERGIZE CIRCUIT
LIFT WIRE (MAY HAVE
TO UNSOLDER).
RE-ENERGIZE AFTER
METER IS IN CIRCUIT

INSTALL METER IS
IN SERIES WITH
CIRCUIT

000.0

FIGURE 4–6
Ammeter shunt

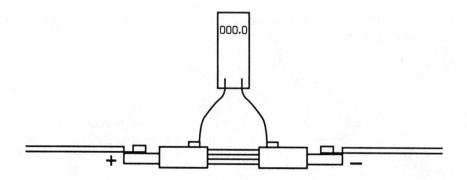

FIGURE 4–7
DC voltmeter and
polarity in a circuit

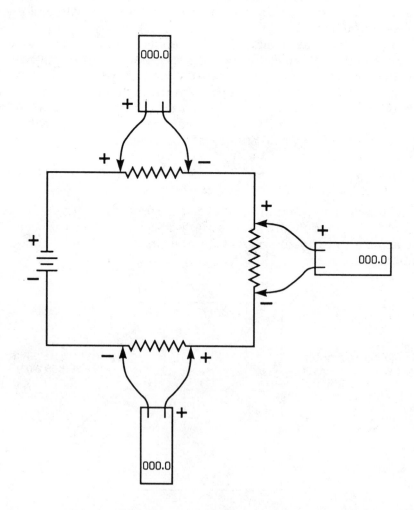

Some typical uses for DC voltmeters include taking voltage readings of batteries. The problem is that so little load is imposed by the meter that often a defective battery will read full voltage. It is not until the battery is loaded down that a problem appears.

DC voltmeters are a good tool for finding blown fuses. Figure 4–8 shows the procedure for determining which of the fuses are bad.

Observe polarity, and make sure DC voltage appears on the line side of the fuses. Take readings from the line side of fuse number two to the load side of fuse one. If voltage is the same as line voltage, then fuse number one is all right. If no voltage is read, fuse one is blown. Next take readings from the line side of fuse one to the load side of fuse two. If voltage is the same as line voltage, the fuse is all right. If no voltage is read, fuse two is blown.

> **WARNING**
> *Treat every circuit as a live circuit.*

OHMMETERS

Ohmmeters are an excellent troubleshooting tool. One of the prime purposes for the ohmmeter is to read resistance in *de-energized* circuits.

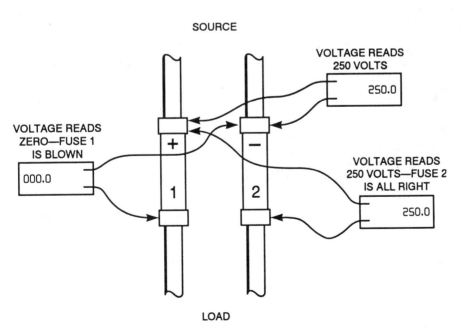

FIGURE 4–8
Checking for bad fuses in DC 250-volt supply line

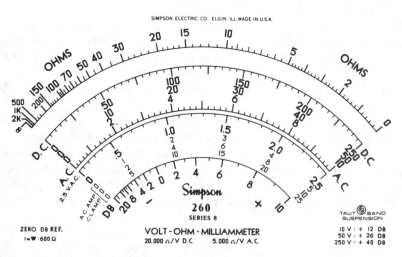

**FIGURE 4–9
Ohmmeter scale.
(Courtesy of
Simpson Electric
Company)**

Some of the more obvious uses for the ohmmeter include testing the condition of out-of-circuit fuses and checking that switches complete a circuit.

Accuracy for analog ohmmeters is given in terms that at first seem confusing. Notice in Figure 4–9 that the ohm scale is not linear. Numbers to the right of center have more distance between them. As the scale approaches the left edge, numbers are crowded and difficult to read. Because of this, accuracy for the ohm scale is usually given as a percentage of arc, or degrees of arc.

What is arc? All circles contain 360°. An arc is part of a circle, and the scale plate of a typical multimeter, as shown in Figure 4–9, is approximately 90° from one end to the other. If accuracy is given as 2 percent of arc, that means that any reading on that dial can be 1.8° high or low (.02 × 90). On the right side of the scale, where values are expanded, reading the meter is not too difficult. However, on the high side (left side), values are cramped together and are difficult enough to read without adding or subtracting an error. In order to read ohms as accurately as possible, select a range that will allow the pointer to be read from the center to the right-hand edge of the meter.

Figures 4–10A and 4–10B show the incorrect and correct methods of taking readings. In order to read resistance, the ohmmeter circuit has an internal battery (or batteries). This voltage through internal resistance and measured resistance equals current measured by the meter element. If connected to an energized circuit, internal resistors designed to handle 1/4 watt of power may be subjected to many times that load. Results are usually predictable. The resistor acts like a fuse and burns up.

For your information, a chart showing resistor markings is shown in Figure 4–11. Some other uses for the ohmmeter include the following.

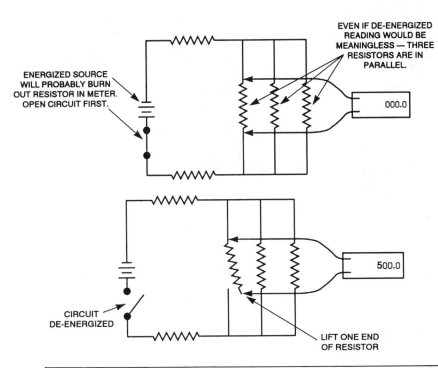

FIGURE 4–10A
Improper use of ohmmeter

ENERGIZED SOURCE WILL PROBABLY BURN OUT RESISTOR IN METER. OPEN CIRCUIT FIRST.

EVEN IF DE-ENERGIZED READING WOULD BE MEANINGLESS — THREE RESISTORS ARE IN PARALLEL.

000.0

FIGURE 4–10B
Proper use of ohmmeter

CIRCUIT DE-ENERGIZED

LIFT ONE END OF RESISTOR

500.0

FIGURE 4–11
Resistor color code chart

BAND COLOR	BAND 1 1ST SIGNIFI-CANT FIGURE	BAND 2 2ND SIGNIFI-CANT FIGURE	BAND 3 MULTIPLIER	BAND 4 TOLERANCE %
BLACK	0	0	$X10^0$	
BROWN	1	1	$X10^1$	
RED	2	2	$X10^2$	
ORANGE	3	3	$X10^3$	
YELLOW	4	4	$X10^4$	
GREEN	5	5	$X10^5$	
BLUE	6	6	$X10^6$	
PURPLE (VIOLET)	7	7	$X10^7$	
GRAY	8	8	$X10^8$	
WHITE	9	9	$X10^9$	
GOLD			X0.1	5
SILVER			X0.01	10
NO COLOR				20

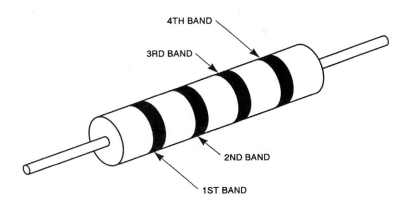

4TH BAND

3RD BAND

2ND BAND

1ST BAND

FIGURE 4–12
Capacitor check

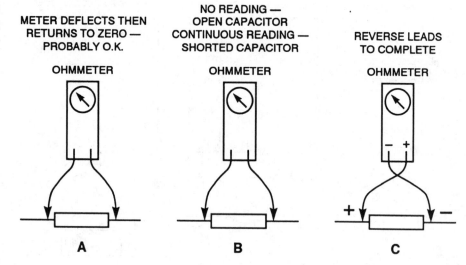

METER DEFLECTS THEN RETURNS TO ZERO — PROBABLY O.K.	NO READING — OPEN CAPACITOR CONTINUOUS READING — SHORTED CAPACITOR	REVERSE LEADS TO COMPLETE
OHMMETER	OHMMETER	OHMMETER
A	B	C

Capacitor Check

An out-of-circuit test for capacitors can be made using the ohmmeter, as shown in Figure 4–12. First be certain that the capacitor voltage rating is higher than that of the ohmmeter battery. Observing capacitor polarity, touch the ohmmeter test probes to the capacitor leads. As the capacitor charges, it temporarily acts like a short circuit, and the analog meter will try to deflect up scale. Once charged, the capacitor will have no current flow, and the ohmmeter will return to zero. Now reverse the leads of the ohmmeter and do the identical test. The same results should be obtained. Remember that on some capacitors the deflection of the needle on the ohmmeter is so quick that it cannot be observed. If the ohmmeter continues to read a high resistance, the capacitor may be shorted. Using an ohmmeter for capacitor testing only checks the resistance of the capacitor and does not check for changes in capacitance or load failures.

Diode Check

Figure 4–13 shows a quick out-of-circuit check for diodes. Be certain that the voltage rating of the diode is higher than that of the ohmmeter battery. Diodes will pass current in only one direction. So, observing polarity as shown, touch the diode leads and observe the reading. A high resistance reading should be observed. Reverse the leads and read

CONNECT THE OHMMETER AS
SHOWN, OBSERVING POLARITY.
A READING SHOULD BE OBTAINED.
IF NO READING DIODE IS OPEN.

REVERSE CONNECTIONS —THE
METER SHOULD NOT SHOW A
DEFLECTION. IF READINGS OCCUR
BOTH WAYS, DIODE IS SHORTED.

FIGURE 4–13
Diode check

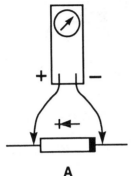

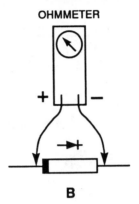

again. This time readings should be near zero. No readings in either direction indicate an open diode. High readings in both directions show a shorted diode.

Primary to Secondary Potential Transformer Fault

Isolation transformers, those typically used for instruments (primary/120 volts), show no physical connection between primary and secondary windings. Before putting them into service, this quick check on de-energized transformers, as shown in Figure 4–14, will show obvious problems.

Using the highest ohmmeter range, place one probe on either primary lead (H_1 or H_2). Place the other probe on either secondary lead (X_1 or X_2). If a steady ohm value is read, try a lower range. If at the lower range the reading still holds, that transformer is definitely suspect.

AC INSTRUMENTS

Before using AC test equipment, it might be a good idea to review the types of systems they will be used on.

Figure 4–15 shows typical transformer connections for single-phase service.

FIGURE 4–14
Checking potential
transformer for
primary to secondary
problem with an
ohmmeter

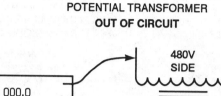

POTENTIAL TRANSFORMER
OUT OF CIRCUIT

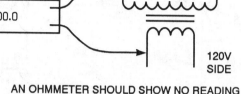

AN OHMMETER SHOULD SHOW NO READING
ON ANY RANGE BETWEEN PRIMARY AND
SECONDARY

Single-Phase, Two-Wire Service

Some really old homes may still have this type of system. The power
company ran one hot lead (circuit A or B) and a neutral to the service
entrance. The maximum voltage in this type of service is 120 volts.

Since the end of World War II, most homes have had single-phase,
three-wire service installed. Two hot leads, A and B, plus a neutral, are
run from the transformer to the customer's service. Either A or B to
neutral will provide 120 volts; A to B will provide 240 volts.

At first glance it might appear that this is a two-phase circuit. Yet
because each hot lead comes off the same transformer coil, the correct
name is single-phase.

FIGURE 4–15
Single-phase
transformer
connection

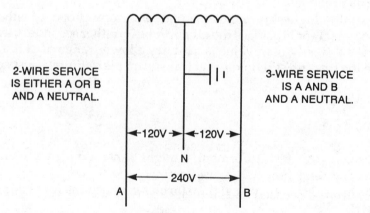

2-WIRE SERVICE
IS EITHER A OR B
AND A NEUTRAL.

3-WIRE SERVICE
IS A AND B
AND A NEUTRAL.

Three-Phase, Three-Wire Service

Three-phase, three-wire service is shown in Figure 4–16. It is primarily used when loads are motors or pumps. Internally, transformers are used to step down voltage for lighting. Voltage between each phase will read the same. Should a reading be taken between any phase and a safety ground, the meter will indicate same low, useless voltage.

Three-Phase, Four-Wire Delta (Δ)

This is a variation on the three-phase, three-wire service. As shown in Figure 4–17, a grounded center-tap connection is added to the transformer in A phase. This provides two 120-volt lighting circuits, in addition to the 240 volts across any two conductors. Using A, B, and C provides three-phase voltage to motors.

The C phase of this system is used for three-phase duty only. A reading taken between C phase and neutral will indicate 208 volts. Some of the names applied to C phase include power leg, delta leg, dog leg, stinger, and red leg.

Three-Phase, Four-Wire Wye (Y)

Three-phase, four-wire, wye service is used both in industry and commercial buildings. As shown in Figure 4–18 (page 96), it has the advantage of providing 120 volts between each phase and neutral. Phase-to-phase readings are 208, so three-phase motors can also be supplied.

In some commercial installations such as apartment houses, two hot leads and a neutral are run from the three-phase, four-wire Y service. This provides 120 volts for lighting. Voltage across the phases is only 208 volts. However, most modern appliances, such as stoves and water heaters, can run on either 208 or 240. While typically called network installations, technically these could be called a two-phase service.

AC AMMETERS

Besides analog and digital meters, two other groups of AC ammeters must be discussed. They are in-line and clamp-on ammeters. In-line AC ammeters, especially of the multi-tester variety, are not very versatile unless combined with an instrument CT, or AC clamp-on probe. These devices were discussed in Chapters 2 and 3, should you need a refresher.

> **WARNING**
> *Treat every circuit as a live circuit.*

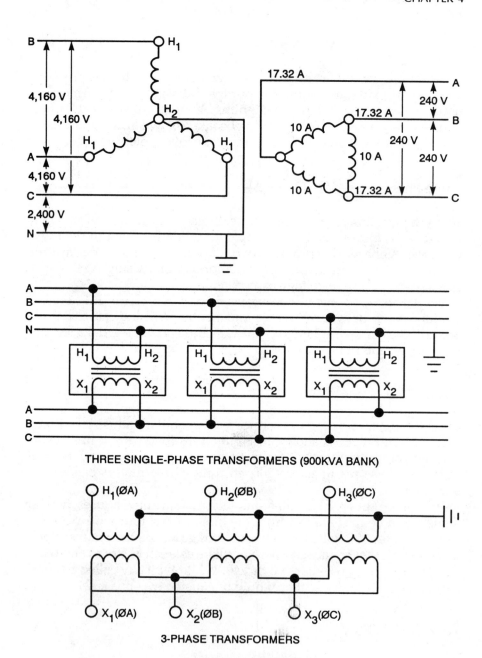

THREE SINGLE-PHASE TRANSFORMERS (900KVA BANK)

3-PHASE TRANSFORMERS

On 3Ø 3-wire service phase and line voltage are the same.
Line current is 1.732 x phase current.

FIGURE 4–16
Three-phase, three-wire

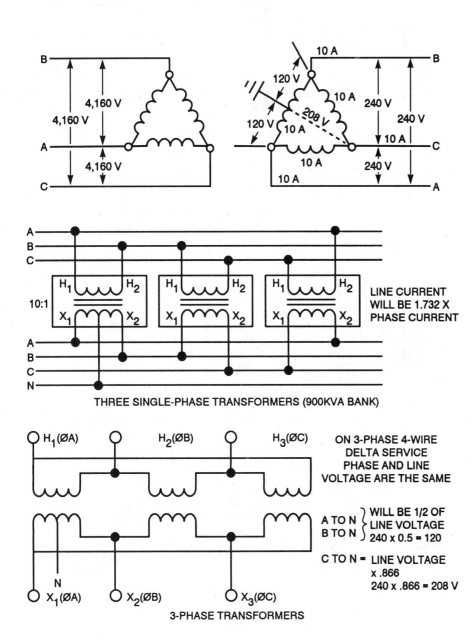

FIGURE 4-17
Three-phase, four-wire delta

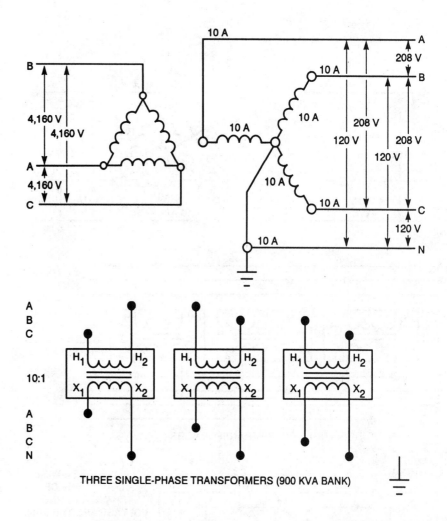

THREE SINGLE-PHASE TRANSFORMERS (900 KVA BANK)

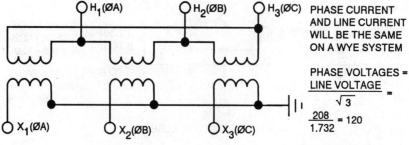

900 KVA 3-PHASE TRANSFORMERS

PHASE CURRENT AND LINE CURRENT WILL BE THE SAME ON A WYE SYSTEM

PHASE VOLTAGES = $\dfrac{\text{LINE VOLTAGE}}{\sqrt{3}}$ =

$\dfrac{208}{1.732}$ = 120

LINE VOLTAGE = PHASE VOLTAGE x $\sqrt{3}$ = 120 x 1.732 = 208

FIGURE 4–18
Three-phase, four-wire wye

Remember the caution about CTs in Chapter 2! As current flows through the primary of a current transformer, it sets up a magnetic field in the core and is induced in the secondary. As long as there is a path across the terminals of the secondary, either a shorting bar or the coil of an ammeter, the current induced in the secondary will be at a ratio equal to the current rating of the CT.

If, however, a meter lead is suddenly pulled off the secondary coil, all the magnetic flux produced by the primary coil will be diverted to trying to keep current flowing in the secondary, which is now open circuited. As a result, a high voltage, called crest voltage, develops across the secondary. For a few cycles, this voltage can be as high as the primary voltage the CT is installed on.

When taking readings from a CT the safest approach is to short out the secondary using the built-in device. Use spade lugs to connect the meter to the secondary terminals, as alligator clips have a nasty habit of slipping off and creating a hazardous situation. Once the CT is installed in the circuit and the meter connected, remove the short across the secondary. When readings are taken, again short out the CT secondary and replace the original circuit.

> *WARNING*
> *A basic rule for using current transformers is never open the secondary of a current transformer under load.*

CLAMP-ON AMMETERS

A clamp-on ammeter is one of the handiest tools for the maintenance electrician. Low cost and versatility make it a must for every electrical worker. As discussed in Chapter 3, analog units are available with switch selection that can read out amperes in peak, rms, or average current. Wide jawed units, as shown in Figure 4–19, allow reading from bus bars or the bodies of large fuses. Some of the more obvious uses include the following.

Checking for Open Phase in a Three-Phase Feeder Circuit

If one phase of a three-phase feeder circuit opens, a three-phase motor may continue to run under a condition called single-phasing. The usual symptoms include low torque, overheating, and failure to start.

If the motor is still running when the technician arrives, a quick check on each phase with the clamp-on ammeter will tell which is at fault. Once the motor is shut down, check the fuse, cable, and connections of the problem phase.

Testing Centrifugal Switch

Single-phase motors often have a capacitive circuit to help the motor to start. Once the motor is up to speed, a centrifugal switch opens the

FIGURE 4–19
Large jaw digital ammeter. (Courtesy of HD Electric Company, Deerfield, IL, U.S.A.)

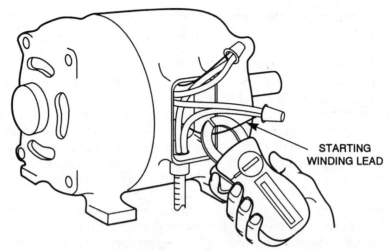

STARTING
WINDING LEAD

CURRENT INDICATION AFTER MOTOR IS UP TO SPEED
MEANS CENTRIFUGAL SWITCH DID NOT OPEN

circuit. If a problem is suspected, a clamp-on ammeter will tell if current continues to flow in the capacitive circuit, as shown in Figure 4–20.

AC VOLTMETERS

The AC voltmeter is another versatile tool that comes in a variety of models, the multimeter being the most popular. Typical voltage ranges are from just a few volts to 750 volts AC.

Should readings above 600 volts be required, a step-down transformer is usually used. These transformers are available in voltages that range up to thousands of volts on the primary side, and step-down voltage to 120 volts in the secondary, as shown in Figure 4–21. Meters used with potential transformers are usually 150 volts full scale to permit some over voltage capability.

Some of the typical uses for voltmeters are for checking out fuses, as shown in the section under DC voltmeters. Other uses are as follows.

Checking the Ratio of a Current Transformer

Using voltage is a simple way of verifying the ratio of an out-of-circuit current transformer. Figure 4–22 shows the test set up.

Calculate the ratio of the CT. In the example shown, 200/5 equals 40:1. This means that for every 40 volts injected into the secondary, 1

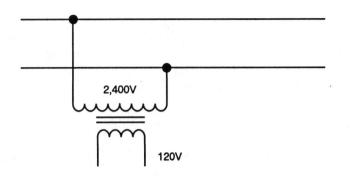

FIGURE 4–21
Potential transformer

2,400V

120V

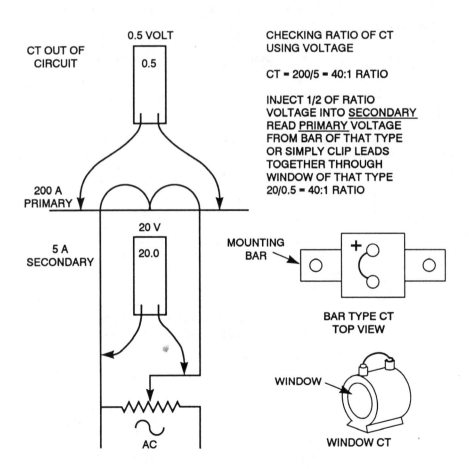

FIGURE 4–22
Checking current transformer ratio using voltage

CT OUT OF
CIRCUIT

0.5 VOLT

0.5

200 A
PRIMARY

5 A
SECONDARY

20 V

20.0

AC

CHECKING RATIO OF CT
USING VOLTAGE

CT = 200/5 = 40:1 RATIO

INJECT 1/2 OF RATIO
VOLTAGE INTO SECONDARY
READ PRIMARY VOLTAGE
FROM BAR OF THAT TYPE
OR SIMPLY CLIP LEADS
TOGETHER THROUGH
WINDOW OF THAT TYPE
20/0.5 = 40:1 RATIO

MOUNTING
BAR

+

BAR TYPE CT
TOP VIEW

WINDOW

WINDOW CT

volt will appear in the primary. For reasons of safety, and in order not to stress the insulation, inject just one-half of this value into the secondary (40 × 0.5 = 20 volts). Read 1/2 volt by connecting voltmeter leads to the CT bus bar. If the unit under test is a window type, pass the leads through the window and clip them together.

Checking the Ratio of a Voltage Transformer

In checking the ratio of an out-of-circuit VT (or PT for potential transformer), again the ratio must first be calculated. In the example and test setup shown in Figure 4–23, a 480/120 volt transformer is shown. This results in a 4:1 ratio. There is no need to inject 480 volts into the primary to test the unit. Pick some low voltage, 40 volts, for example; inject that voltage into the primary and read 10 volts on the secondary (40/10 = 4:1). Never inject test voltage into the secondary, as the transformer will step up as well as step down voltage.

The voltmeter is also a useful tool for troubleshooting motors. Figure 4–24 shows a series of tests that require energizing the motor circuit.

FIGURE 4–23
Checking potential
transformer ratio

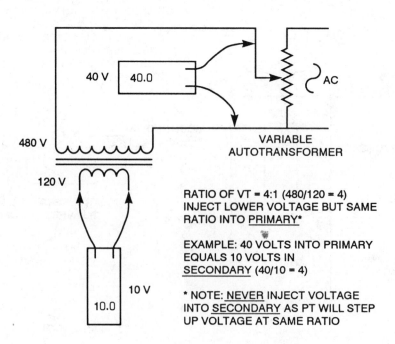

40 V   40.0

AC

VARIABLE
AUTOTRANSFORMER

480 V

120 V

RATIO OF VT = 4:1 (480/120 = 4)
INJECT LOWER VOLTAGE BUT SAME
RATIO INTO <u>PRIMARY</u>*

EXAMPLE: 40 VOLTS INTO PRIMARY
EQUALS 10 VOLTS IN
<u>SECONDARY</u> (40/10 = 4)

10 V   * NOTE: <u>NEVER</u> INJECT VOLTAGE
INTO <u>SECONDARY</u> AS PT WILL STEP
10.0   UP VOLTAGE AT SAME RATIO

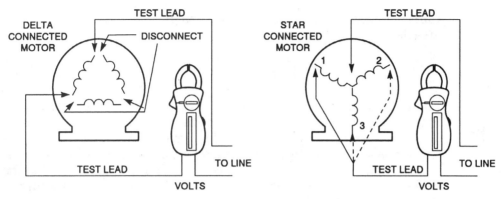

WINDING IS OPEN IF THERE IS NO VOLTAGE INDICATION ACROSS THE WINDING.

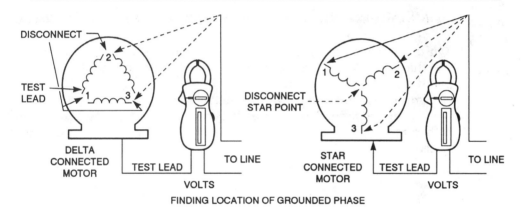

FINDING LOCATION OF GROUNDED PHASE

GROUNDED PHASE IS INDICATED BY A FULL LINE VOLTAGE READING.

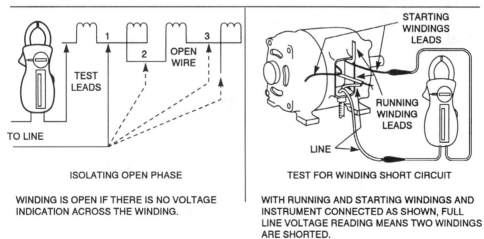

ISOLATING OPEN PHASE

WINDING IS OPEN IF THERE IS NO VOLTAGE INDICATION ACROSS THE WINDING.

TEST FOR WINDING SHORT CIRCUIT

WITH RUNNING AND STARTING WINDINGS AND INSTRUMENT CONNECTED AS SHOWN, FULL LINE VOLTAGE READING MEANS TWO WINDINGS ARE SHORTED.

FIGURE 4–24
Using voltmeter to check motor circuits. (Courtesy of Amprobe Instruments®)

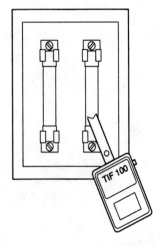

**FIGURE 4–25
Checking fuses.
(Courtesy of TIF
Instruments, Inc.)**

Carefully measure the points indicated to pinpoint where the problem area lies.

AC VOLTAGE DETECTORS

Figure 4–25 shows how a voltage detector can be used to check fuses. Starting above the fuse, make certain the circuit is energized. Tracing over the fuse to the conductor below, make certain that voltage is getting past the fuse. Repeat with each phase to locate the blown fuses.

Although in Figure 4–26 the voltage detector is being used to find the broken point in a heating element, the same method will work with any AC carrying wire.

SUMMARY

Quite often test equipment must be operated in a hostile environment, yet the chief cause of test equipment failure is careless application and handling by the test technician.

Simply remembering that ammeters are connected in series, and voltmeters in parallel, will lessen the chances of damaging equipment. When reading resistance, measurements should be made on de-energized circuits.

**FIGURE 4–26
Checking heating
element. (Courtesy
of TIF Instruments,
Inc.)**

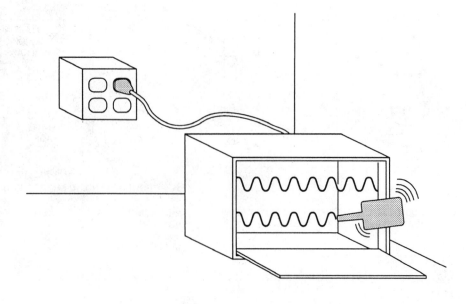

Some equipment is designed to be operated in a wet environment, but most is not. Moisture, dust, and grime are all enemies of meter accuracy.

Calibration stickers serve a purpose of telling when equipment was last tested, and also when test equipment is again due for test.

Meter accuracy may be defined as percent of readings, such as ±1 of any reading on a given scale. Or, accuracy may be percent of full scale, for example ±2% of full scale. An example might be ±2% of a 150-volt scale. Errors could then be ±3 volts on any reading on the scale.

Shunts can be used to extend the ranges of DC ammeters. On AC meters, current or voltage transformers are used to extend the range of ammeters and voltmeters.

Meter sensitivity determines the amount of voltage or current necessary for the measuring circuit to react. The more voltage or current that must be drawn from the measured circuit, the less accurate the measurement of that circuit.

Adjusting the meter zero position is the first step to ensuring meter accuracy. Not only should the pointer indicate zero, but the adjusting mechanism should not bind the meter movement.

Parallax is the condition that exists when a technician views the analog pointer of a meter from the side. Since any angle will cause a reading error, manufacturers force the view to be directly above by using a knife-edged pointer, a mirrored scale, or both.

REVIEW QUESTIONS

1. How close can trained personnel approach a 2,000 volt overhead line?

2. What is the AC test voltage for class 0 protective gloves?

3. What is the primary cause of damaged meter test equipment?

4. What two dates are included on a calibration sticker?

5. A shunt is one device that can extend the range of a DC ammeter. What is another?

6. Explain what percent of arc means in analog ohmmeters.

7. What is the difference between three-phase, three-wire and a three-phase, four-wire delta service?

8. What is the primary rule when working with current transformers?

9. In testing the ratio of a current transformer, why is the voltage injected less than the actual ratio?

10. In ratioing a step down voltage transformer, why isn't the voltage injected into the secondary?

Troubleshooting and Testing

The transformer shown in Figure 4–15 only had the A to Neutral connected across a number of 120-volt loads. A work order was issued to install a B conductor from the transformer so that 240 volts across A and B conductors are available for future loads as well as to feed the potential coil of a monitoring meter. Loads originally all on A were to be split up so that circuits A and B would be equally loaded.

Work was interrupted in the middle of the conversion, and another crew finished the job late the next day. A quick check showed that all 120-volt circuits were operational. Since the 240-volt meter showed no indication, the crew reported a defective meter and left for the day.

In troubleshooting the installation, a test technician found the meter to be operational and reading 120 volts when connected from A to neutral or B to neutral. The reading from A to B again showed zero volts.

QUESTION

What was the probable cause for the voltage readings the test technician found?

ANSWER

This situation is typical when a second crew, not completely understanding the work to be done, takes over from one that does. Arriving at the scene, they found that loads had been split between the old conductor A and the new conductor B.

They reasoned that the purpose of the job was to cut the ampacity load on A conductor, so they connected the new conductor to A as well.

Voltage readings from the old to the new conductor (both A) will read zero volts, but 120 volts to neutral. By moving the new conductor from A terminal to B terminal of the transformer all voltages will read correctly.

Test Bench Lab
Test Equipment

5

Introduction

Past chapters have covered how meters measure voltage, amperage, and resistance. The instruments discussed thus far have primarily been of the panel or portable types. In this chapter, the type of equipment used in the repair shop will be covered. Usually these instruments are more accurate, or it may be the type of cases these devices are mounted in, such as the multimeter shown in Figure 5–1.

Variable autotransformers provide a range of controlled AC voltage, used for either meter calibration or as a regulated input voltage to a device under test.

A DC bench power supply is similar to the AC variac, except it is exclusively a source for DC voltage. Filtering within the test set assures that DC ripple is held to a minimum.

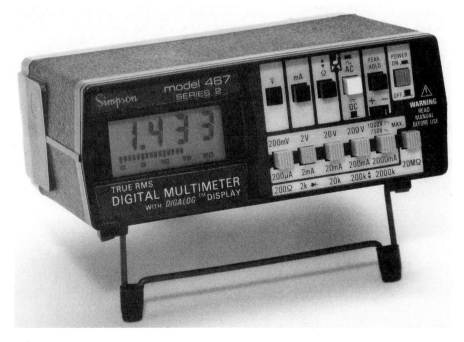

**FIGURE 5–1
Bench top digital multimeter.
(Courtesy of Simpson Electric Company)**

Decade resistance boxes allow the test technician to select values of resistance in very small increments for either instrument calibration, or for selecting appropriate values in equipment design.

When precise measurement of resistance is necessary, the Wheatstone bridge is the one of the best instruments. Uses for the bridge include testing of resistance temperature devices and pinpointing the location of trouble spots in control or power cables.

Transducer calibrators are used to test heat sensing instruments in the industrial environment. A single unit is capable of testing transducers, recorders, indicators, or controllers.

As more equipment includes integrated circuits, logic pulsers and probes become more important tools in troubleshooting. Pulsers and probes enable the test technician to follow the logic flow through the various circuits, and to pinpoint inoperative components.

OBJECTIVES

After studying this chapter, the student should be able to:

- *Explain how an autotransformer controls AC voltage output.*
- *Identify two types of DC bench power supplies.*
- *Explain how a decade resistance box operates.*
- *Understand how the Wheatstone bridge measures resistance.*
- *Identify uses for a thermocouple RTD tester.*
- *Explain how a logic pulser operates.*
- *Explain how the logic probe indicates when an integrated circuit is high or low.*

VARIABLE AUTOTRANSFORMERS

Transformers can be wound with one primary input voltage feeding a coil, and a number of outputs on that same coil. This allows the user a number of voltage outputs, depending on his needs.

In testing, it is often necessary to have a continuously variable voltage available. One of the devices most often used is the variable autotransformer, such as the one shown in Figure 5–2. This device has a movable brush-tap point, riding on the winding of the transformer. In some models, rotation of the brush will deliver a variable output voltage from zero to 15 percent above the input line voltage. The overrange allows the user to compensate for low line input voltage, as well as to offer more flexibility. Figures 5-3A and 5-3B give an idea of how this is

FIGURE 5–2
**Auto transformer.
(Courtesy of Staco
Energy Products
Company)**

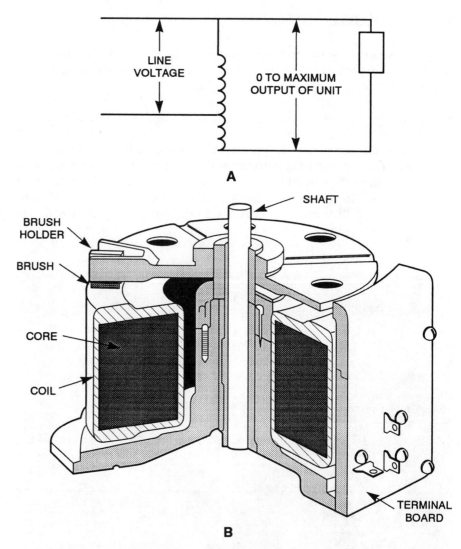

FIGURE 5–3
Variable auto
transformer

done. Notice from the diagram that the autotransformer is not an isolation transformer. The device is nothing more than a continuous coil, which acts much like a voltage divider circuit.

The linear dial of the variac, marked from 0 to 100 percent of maximum output voltage, is merely a guide. Line voltage supplying the unit can vary continuously, so a 50 percent setting would only mean that output is approximately one half of whatever value input voltage is at the moment.

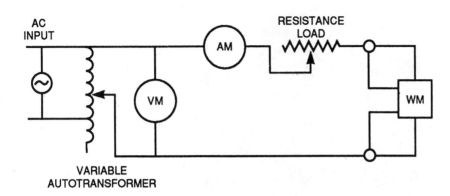

FIGURE 5–4
Wattmeter check
with variable
autotransformer

Variable transformers are offered in single- and three-phase, manual or motor-driven, units. Many operate from 120 volts, but can be ordered for operation on 240 or 480 volts. Even a choice of clockwise or counterclockwise rotation is available in some models.

The variable autotransformer has many uses in the electrical maintenance shop. In addition to checking the accuracy of voltmeters, the variac is often used to control watts for testing wattmeters. A later chapter will explain wattmeters and kilowatt-hour meters in detail. But you may recall from the chapter on basic electricity that power in an AC or DC resistive circuit equals E x I. If voltage is put through a variable resistor, which in turn controls amps, watts will be developed.

In the test circuit shown in Figure 5–4, 120 volts are fed from the variable transformer through 24 ohms of resistance. By formula, watts will equal

$$\frac{E^2}{R} = 600 \text{ Watts} \qquad \frac{120 \times 120}{24} = \frac{14{,}400}{24} = 600 \text{ Watts}$$

DC BENCH POWER SUPPLY

At times, it will be necessary to determine if printed-circuit boards are good or bad. One of the simplest ways is to replace the board you think might be bad with a good one. However, if no spare is available, a source of regulated DC is a must in troubleshooting the board. Figure 5–5 shows a typical unit.

An ideal regulated DC bench power supply maintains a constant DC output, despite changes in external load requirements or changes in input voltage. Not only must the unit provide DC voltage, but the ripple associated with rectified DC must be kept to a minimum. Stan-

FIGURE 5–5
DC power supply
unit. (Courtesy of
B + K Precision)

dard, general-purpose regulated power supplies are used in the laboratory as shown in Figure 5–6.

DC power supplies may be divided into two functional categories: constant-current and constant-voltage.

FIGURE 5–6
DC power supply

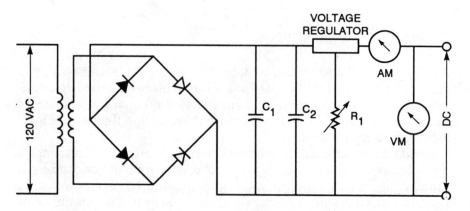

Constant-current power supplies maintain a constant current through a load resistor, regardless of the value of the load resistance, as long as it stays within a certain voltage range. Constant-voltage power supplies maintain a fixed voltage across the load resistor, as long as the current through the load resistor stays within a specified range.

Connection

For proper operation, the power supply is usually positioned so that it is level, and that all louvers of the case are unobstructed. The power supply cable connections may differ with intended uses, and the input and output connections must be verified before the unit is placed into operation.

Most bench power supplies have a permanently installed AC input voltage plug, but, some larger three-phase units do not, and must be cabled directly to the receptacle instrument panel marked "three-phase AC input" and power source. The ground strap is connected between the ground connection on the power supply and an earth ground.

The DC bench power supply will have one or more outlets on the front panel to supply a DC output. Connections made from the power supply to the unit are determined by the application, with the positive and negative leads from the power supply connected to the corresponding connections on the unit to be supplied.

Measurement

Bench power supplies normally have a DC voltmeter and an ammeter mounted on the front panel. Output voltage is monitored by the voltmeter that is connected across the output circuit, while output current is read by operating the ammeter switch that is connected across the ammeter to prevent overloading if output current exceeds the rated output. The potentiometer, and the ammeter in series with it, are connected across an external shunt in the output circuit.

DECADE RESISTANCE BOX

A decade resistance box provides a source of precision resistors over a given range. Any resistance, from the lowest to the highest values on the unit, can be dialed into a circuit.

Application

Uses for the resistance box include testing the accuracy of ohmmeters and finding the value of a burned-out resistor and unidentified value of a resistor in a meter circuit. Resistance is simply added while applying voltage to the test leads until the meter reads the correct value applied.

Connection

Two binding posts are provided on the front panel of the resistance box. In some units a third terminal provides a ground connection to the metal case and forms a shield for the instrument. Resistance between the two binding posts corresponds to the setting of the dials, which are arranged to read resistance directly in ohms. In the decade box, all resistances are arranged in units of ten, and several decade units can be connected in series to make a wide range of resistance values available.

Measurement

The decade resistance box shown in Figure 5–7A has nine equal resistors connected in series between the contact points of a ten-position rotary switch. Each dial has a descending order of values, from 100 ohms on the highest dial to 0.01 ohms on the lowest dial. As shown in Figure 5–7B, by selecting the value needed, all ohms from zero to 999.99 ohms can be selected in 0.01 ohm steps.

Care must be taken when using the resistance box in any live circuit. Each of the selector dials has a maximum current value that the resistors can handle. For example, high value resistors (100 ohms) can only handle 50 milliamperes (0.050), while the one (1) ohm resistors can handle up to 5 amperes. Each dial can be turned to allow rapid and direct change of settings from 0 to 9.

Decade resistance units are available with values from fractions of an ohm, or megohms.

Wheatstone Bridge

Figure 5–8 shows a typical Wheatstone bridge. A basic bridge circuit for resistance measurement is shown in Figure 5–9. The circuit has a power source, two fixed resistors (RA and RB), a variable resistor (R), and a galvanometer. RX is the unknown resistance.

FIGURE 5–7A
Decade resistance
box. (Courtesy of
Shalltronix
Corporation)

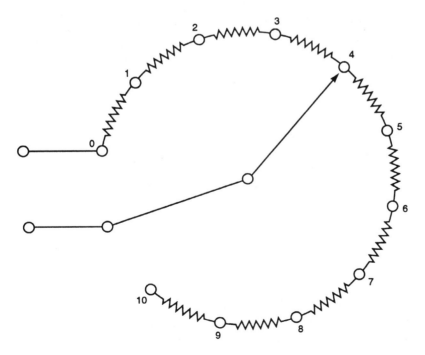

FIGURE 5–7B
Decade resistance
box selector switch

FIGURE 5–8
Wheatstone bridge.
(Courtesy of
Shalltronix
Corporation)

FIGURE 5–9
Wheatstone bridge
circuit

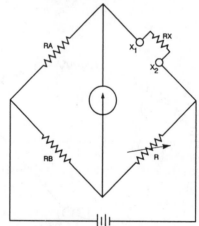

To read resistance
RX = A × R
where
RX = Resistance
A = Setting of Multiply by dial
R = Settings of rheostat dials

Example

$A = \frac{1}{10}$ (0.1) Set R V M switch to R (resistance)

R = 9.761

RX = 0.1 × 9761 = 976.1 Ω

RA and RB are both controlled by the ratio arm (MULTIPLY BY dial). R is made up of four rheostat dials. Total R is read and then multiplied by the MULTIPLY BY dial setting for the final resistance readings.

Operation of the bridge circuit is based on the fact that if the potentials at points e and f are equal, no current will flow through the galvanometer. This condition only occurs if the ratio of RA to RB equals the ratio of RX to R.

$$\text{In simpler form: } \frac{RX}{RB} = \frac{RX}{R}$$

For measurement, the unknown resistance is placed in the circuit, and the MULTIPLY BY and R dials are adjusted until the galvanometer is nulled. Three buttons, labeled GA SENS .01, .1, and 1 control current flow to the galvanometer. Start with the least sensitive, .01, and continue balancing until a null reading is obtained with the 1 button depressed. If the galvanometer deflects toward the end marked "+," there is too little resistance. If it deflects toward the "–," too much resistance has been added. R is found by reading the four rheostat dials and multiplying this value by the setting on the MULTIPLY BY dial.

The method above satisfies the following formula:

$$RX = \frac{RA}{RB} = R$$

To avoid the need to move leads during tests, bridges have an RVM switch that changes lead or galvanometer positions in the basic circuit. R stands for resistance, and will be used for most basic tests. V is for Varley, a circuit variation that allows readings to be taken through the ground terminal. M is for Murray, a circuit that changes resistor relationships and galvanometer routing within the bridge.

Switches can also select between internal batteries, or galvanometers, or external units.

Applications

There are many applications for the Wheatstone bridge and many variations of it. A Wheatstone bridge is often used to measure resistance temperature detectors (RTDs). RTDs perform a wide range of duties in industry. These include monitoring chemical processes in pharmaceutical labs to measuring the temperature of large industrial motors. Resistance changes with temperature. By reading the resis-

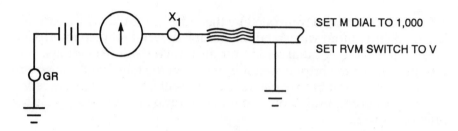

FIGURE 5–10
Identifying
grounded conductor

SET M DIAL TO 1,000

SET RVM SWITCH TO V

tance with the unit in ice water, and again in boiling water, and comparing results with the manufacturer's chart, accuracy can be determined.

Another application for the Wheatstone bridge is finding the location of short circuits or grounds in control cables or de-energized, short, submarine, power cables, Figure 5–10. One method of how the location of a ground on a de-energized conductor can be located using the Murray loop is shown in Figure 5–11.

1. Set the MULTIPLY BY dial to 1,000.
2. Set the GA (Galvanometer switch) to RVM position. Unlock the mechanical galvanometer device.
3. Set BA (battery) switch to internal.

FIGURE 5–11
Murray loop test

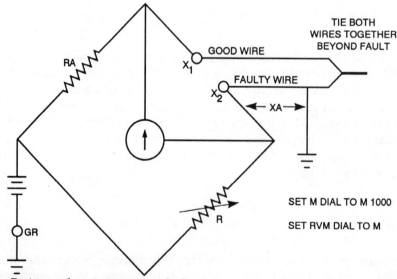

XA = Resistance from instrument to fault
R = Sum of settings of four rheostat dials
A = M settings of multiply by dial
r = Total resistance of good and faulted conductor loop

$$XA = \frac{R \times r}{A + R}$$

4. Set RVM switch to V (Varley) position.
5. Connect a good ground to the GR binding post of the Wheatstone bridge. Identify the grounded wire, connecting each wire to terminal X1 while depressing the .01 SENS button. The faulty wire will be identified when there is a sharp deflection of the galvanometer, as shown in Figure 5–10.
6. Join the faulty wire to a known good wire at a point beyond the fault.
7. Connect the good wire to X1, the grounded wire to X2, and a good ground to GR.
8. Set the RVM switch to M.
9. Add resistance until a null condition exists when SENS button 1 is pushed. Read and calculate this resistance, which is "R" in the formula below. Resistance reading is the total of the good wire, plus the return portion of the bad wire, until it meets the grounded point.
10. Set the RVM switch to the RES position, and add resistance, until a null indication is obtained. Read and calculate the resistance. This reading equals the total resistance of the loop, made up of the good and bad wire.

The following formula can be used to calculate Xa, the distance from the test set to the fault.

$$Xa = \frac{Rr}{A + R}$$

Where:
Xa = Resistance of faulty wire from instrument to the fault
A = Setting of MULTIPLY BY dial
R = Sum of four rheostat dials
r = Total resistance of loop

Once the ohms to the fault have been calculated, a chart such as shown in Table 5-1 can be used to calculate actual distance to Xa.

Lead length and resistance will be a part of the overall readings with a Wheatstone bridge. However, for most situations this can easily be taken into account.

Variations of the basic Wheatstone bridge also can be used for capacitance and inductance measurement. Figures 5-12A and 5-12B show instruments that are available to measure capacitance, inductance, or resistance alone or in combination as impedance.

Table 5–1   Annealed Copper Wire Table (Courtesy of Shalltronix Corporation)

GAUGE NO.	FEET PER OHM		
	32°F	68°F	100°F
12	683.5	641.8	599.2
14	429.7	396.0	370.2
16	270.3	249.1	232.8
18	170.0	156.6	146.4
19	134.8	124.2	116.1
20	106.9	98.52	92.09
21	84.77	78.13	73.03
22	67.24	61.95	57.90
24	42.26	38.95	36.41

For wire other than copper, see note below.

NOTE: Drawn copper—multiply values in Table 5–1 by .973.
Aluminum—multiply values in Table 5–1 by .6
Iron, steel, copper-clad, or similar coated wires, use manufacturer's information.

FIGURE 5–12
Capacitance and inductance measurement

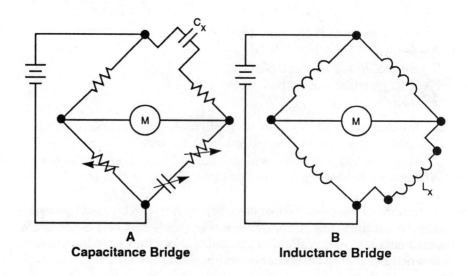

A
Capacitance Bridge

B
Inductance Bridge

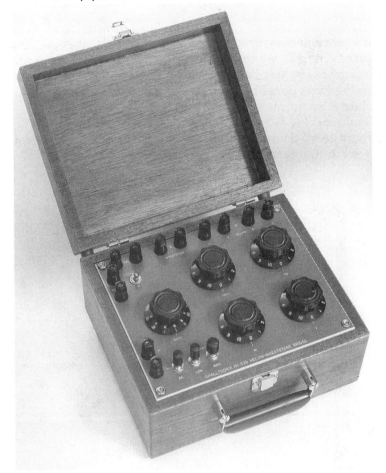

FIGURE 5–13
Kelvin bridge.
(Courtesy of
Shalltronix
Corporation)

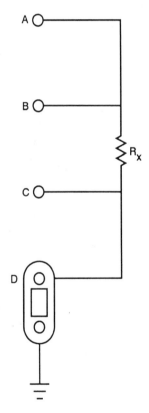

KELVIN BRIDGE

The Kelvin bridge, also known as the double bridge or Thompson bridge, is used for precise measurement of small resistance values down to a fraction of an ohm. The Kelvin bridge as shown in Figure 5–13 has an advantage over the Wheatstone bridge in that the errors caused by lead length and contact resistance at connecting posts and terminals can be eliminated from the calculation.

Connection

There are four bridge terminals in the Kelvin bridge, each of which must be connected. The unknown resistance (RX) is connected to all four terminals, which are designated A, B, C, and D in Figure 5–14. Connections to terminals A and B are interchangeable, as are C and D.

FIGURE 5–14
Kelvin bridge
connections

The resistance is measured between the junction of A and B and the junction of C and D. As shown in Figure 5–14, it is common practice to connect one of the terminals (in this case, terminal D) to a grounding post on the front panel by means of a metal link. This link is removed for certain measurements.

Several possible methods of connection are shown in Figure 5–15. These may be used with a Kelvin bridge or almost any other instrument that uses binding posts for connections.

**FIGURE 5–15
Binding post
connections**

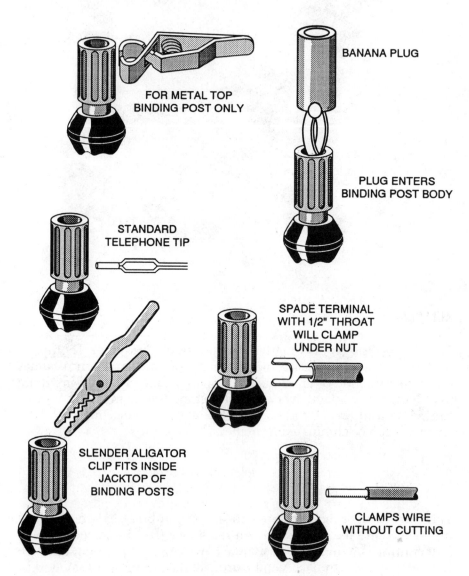

FOR METAL TOP
BINDING POST ONLY

BANANA PLUG

PLUG ENTERS
BINDING POST BODY

STANDARD
TELEPHONE TIP

SPADE TERMINAL
WITH 1/2" THROAT
WILL CLAMP
UNDER NUT

SLENDER ALIGATOR
CLIP FITS INSIDE
JACKTOP OF
BINDING POSTS

CLAMPS WIRE
WITHOUT CUTTING

Measurement

Resistance measurement using a Kelvin bridge is similar to resistance measurement using a Wheatstone bridge. The vendor instruction manual for the particular instrument should be consulted. In general, the first step is to mechanically zero the galvanometer. The measurement is then performed using knobs to adjust each decade of resistance, in descending order, until the galvanometer is balanced.

THERMOCOUPLE AND INSTRUMENT CALIBRATION

Earlier in this chapter it was shown how resistance temperature detectors (RTDs), used for a variety of temperature monitoring tasks in industry, could be tested using a Wheatstone bridge. But RTDs are just one type of heat sensing device. Industry also uses thermocouples, recorders, indicators, and controllers to monitor and modify the temperature of gasses, liquids, or the internal temperatures of such equipment as plastic molding machinery. If there are no instrument technicians on board, electrical maintenance personnel may have to test them all.

Fortunately, equipment is available that makes testing these devices relatively simple. Figure 5–16 shows such a microprocessor controlled calibrator that can be used in the following applications:

- Measuring temperature with thermocouples.
- Comparison of working and reference thermocouples.
- Measuring millivolts from thermocouples or other sources.
- Calibrating thermocouple- or millivolt-type recorders, controllers, and indicators.

Controls and terminals for the unit in Figure 5–16 are as follows:

1. MODE selector. This rotary switch has four mode selections:

   - Off—Turns instrument on or off.
   - Measure—In this position the unit will measure and display temperature (or emf) in accordance with the range switch for the connections made to the selected TC binding post pair (A or B).
   - Output—In this position the selected binding post pair (A or B) is driven at a voltage determined by the setting of the output adjust control. The instrument displays the simulated temperature (or mV) in accordance with the setting of the range selector switch.

FIGURE 5–16
RTD calibrator.
(Courtesy of AVO
Biddle Instruments)

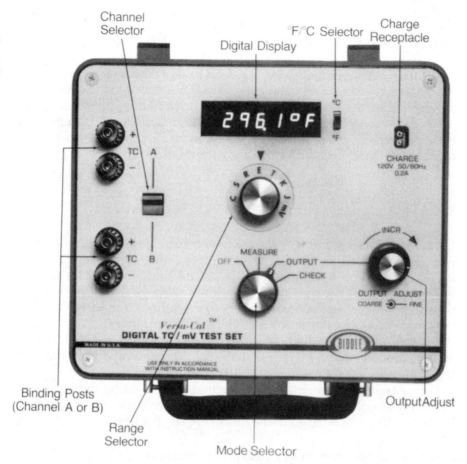

Channel Selector · Digital Display · °F/°C Selector · Charge Receptacle · CHARGE 120V 50/60Hz 0.2A · INCR · OUTPUT ADJUST COARSE — FINE · MEASURE · OFF · OUTPUT · CHECK · Versa-Cal™ DIGITAL TC/mV TEST SET · BIDDLE · Binding Posts (Channel A or B) · Range Selector · Mode Selector · Output Adjust

- Check—In this position the instrument input is disconnected from the binding posts on the front panel and connected to an internally generated reference voltage (approximately 5mV). Normal readings for specific instruments are recorded on a label in the lid as a verification of instrument accuracy.

2. RANGE selector: Selects the desired thermocouple or millivolt range.
3. F/C selector: Selects the desired Fahrenheit or Celsius temperature display.
4. OUTPUT ADJUST: Concentric, dual, ten-turn controls for adjusting the instrument output voltage when operating in the output mode. The inner ("Coarse") control covers the full output range; the outer ("Fine") control covers about 1 percent of the full range. Turning the controls clockwise causes the output voltage to become more positive.

5. BINDING POSTS (Channel A or Channel B): For connection to input or output.
6. CHANNEL selector: Selects the desired pair of binding posts (A or B).
7. CHARGE receptacle: Used to recharge batteries.

SELECTION OF TEST LEADS

Thermocouples are made by creating a junction between two different metals. When this junction, called the hot junction, is heated, voltages developed across the far end of the two metals will track the temperature rise and fall. In measuring thermocouple output, care must be taken in connecting the device to the test set. If the device is connected directly to the test set terminals, which are made of copper, tiny semi-conductors, mounted within the terminals, sense the ambient temperature and compensate all readings except in the millivolt measuring position. When remote readings are taken, the lead wires from the test set must be of the same materials as the thermocouple tested to avoid creating still another dissimilar junction and producing measurement errors. Copper wires are used to connect the two instruments when using the mV range of the test set.

Temperature Measurement with a Thermocouple

Thermocouples are rated, by type, according to the temperature ranges measured. Types of thermocouples this calibrator will test include:

TYPE	TEMPERATURE
J	–346 to +2,192°F (–210 to +1,200°C)
K	–326 to +2,501°F (–200 to +1,372°C)
T	–337 to +752°F (–205 to +400°C) –405 to –337°F (–243 to –205°C)
E	–389 to +1,832°F (–234 to +1000°C) –422 to –389°F (–252 to –234°C)
R,S	–58 to +3,254°F (-50 to +1,768°C) 32 to 2,192°F (0 to 1,200°C)
C	2,192 to 3,812°F (1,200 to 2,100°C) 3,812 to 4,200°F (2,100 to 2,315°C)

To measure a temperature using a thermocouple do the following:

1. Connect the instrument to the thermocouple, as shown in Figure 5–17.
2. Set the channel selector switch to the position corresponding to the pair of binding posts used.
3. Rotate the range selector switch to the TC position that corresponds to the thermocouple type being measured.
4. Set the F/C slide switch to the position corresponding to the desired temperature display.
5. Rotate the mode selector switch to the MEASURE position.
6. Read the value of the measured temperature on the display.

> NOTE: The cold or reference junction compensation circuitry functions in this mode so the test set reads the correct hot junction temperature directly.

Small electric test furnaces are available to heat thermocouples to known temperatures, which are then compared to calibrator readings. However, most thermocouples can be tested for a go/no-go condition by immersing them in boiling water and reading out the temperature on the calibrator as 212° Fahrenheit (100°C).

Temperature Measurements Using mV Range

1. Connect the test set leads to the thermocouple under test as shown in Figure 5–17.

FIGURE 5–17
Test setup for simultaneous input/output. (Courtesy of AVO Biddle Instruments)

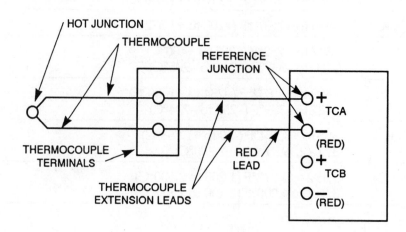

Note: The automatic reference junction compensators (semiconductors located in A and B terminals) are not in the circuit at the mV test position.

2. Measure the thermocouple reference-junction temperature (where the thermocouple is connected to the test set or where the thermocouple is connected to the extension wires coming from the test set) with an accurate mercury-in-glass thermometer, and record the temperature.
3. Convert the reference junction temperature (thermometer) reading to millivolts by referring to an appropriate table provided by test set manufacturers (see the following simplified example).
4. Set the channel selector switch to a position corresponding to the pair of binding posts used.
5. Rotate the range selector switch to the mV position.
6. Rotate the mode selector switch to the MEASURE position.
7. Read and record the value of measured emf.
8. Add together the measured emf reading and the millivolt equivalent of the reference-junction temperature (step 3).
9. Convert the corrected millivolts to temperature directly from the same conversion table. This value is the measuring "hot" junction temperature reading.

Example:

Type J thermocouple immersed in boiling water, 212° Fahrenheit (100° C)

MV reading from (7), above = 4.077 mV

Ambient temperature by thermometer, from (3) above = 74°F

From chart for J thermocouple:

Thermoelectric Voltage in Absolute Millivolts

°F	0	1	2	3	4	5
70	1.076	1.015	1.134	1.162	1.191	1.220

Millivolt reading	4.077
Millivolt correction (at 74°F)	1.191
Sum	5.268

From the same Thermocouple J chart:

Thermoelectric Voltage in Absolute Millivolts

°F	0	1	2	3	4
210	5.207	5.238	5.268	5.298	5.328

Corrected millivolt reading of 5.268 means thermocouple output equals 212°F.

Should the ambient temperature be below 32°F, the correction will be a minus number and subtracted from the millivolt reading.

Calibrating Potentiometer Indicators, Recorders, or Controllers, mV Type

1. Connect the instruments to be checked to the desired binding posts (A or B) of the test set using copper wire and observing proper polarity.
2. Zero the instrument to be checked if required.
3. Set the channel selector switch to the channel in use.
4. Rotate the range selector switch to the mV position.
5. Rotate the mode selector switch to the OUTPUT position.
6. Adjust the OUTPUT ADJUST of the test set so the instrument to be checked balances at the selected check point.

LOGIC PROBE APPLICATIONS

One of the inexpensive aids to troubleshooting printed circuit boards is the logic probe shown in Figure 5–18 and its companion device, the pulsing probe, Figure 5–19.

Logic probes use a number of LEDs to indicate whether a point in a digital signal path is at logic 1 (on, or high), logic 0 (off, or low), or is reading a stream of pulses. All lamps on OFF indicate the absence of any logic signals. Figure 5–20 shows typical circuitry.

Connection

Attach the long clip leads to the 4.7- to 15-volt power supply. Connect the black lead to the negative terminal of the power supply. Connect the red lead to the positive terminal of the power supply. The short black ground lead should be attached to a ground near the point to be tested, to ensure exact performance of the logic probe.

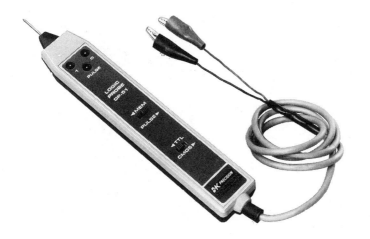

FIGURE 5–18
Logic probe.
(Courtesy of B+K
Precision)

FIGURE 5–19
Pulsing device.
Courtesy of B+K
Precision)

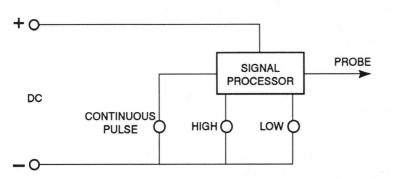

FIGURE 5–20
Logic probe circuit

FIGURE 5–21
Logic pulser

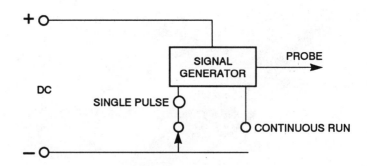

Measurement

Use the following steps to test the logic circuit with the logic probe:

1. Set the DEVICE TYPE switch to CMOS or TTL/LS.
2. Set the MODE switch to the NORMAL mode; the unit will indicate the DC level.
3. Set the MODE switch to the PULSE mode; the logic probe will detect the logic level of the pulse train.

Pulsing devices generally have two positions, as shown in Figure 5–21. The first position allows the user to inject a single pulse into the logic circuit. Position two injects a steady stream of pulses at some predetermined frequency. By using the pulsing device with a logic probe, the chain of a logic circuit can be traced from beginning to end.

SUMMARY

The variable autotransformer allows the operator to select voltages ranging from zero to approximately 115 percent of line voltage, in some models, to control test procedures.

A DC bench power supply, available as either constant voltage or constant-current output, is a useful device as a troubleshooting tool, or as an output used to calibrate measuring instruments.

Decade resistance boxes allow any number of ohmic values to be selected for instrument calibration, as well as equipment design.

The Wheatstone bridge uses a galvanometer as part of its measuring circuit. Once the resistance in the arms of the bridge and the unknown resistances are balanced, the meter is nulled. Resistance is then read directly from the four rheostat dials.

A thermocouple tester is a portable device that is used to measure thermocouple output and calibrate instruments. It can simulate or measure millivolts as an input or output.

Pulsing devices and logic probes are an inexpensive means for troubleshooting printed circuit boards.

Review Questions

1. What is the relationship between the dial markings of a variac and the actual voltage output?

2. Why is there no isolation between the input and output of a variable transformer?

3. What are two types of DC bench power supplies?

4. What is the main difference between DC voltage when rectified by a full wave bridge, and the output of a DC power supply?

5. Do decade resistance boxes operate by combining resistances in parallel or in series to obtain the values desired?

6. When using the Varley position on the RVM switch, how does the voltage return to the test set?

7. Since the Wheatstone bridge reads one resistance at a time, how can it test a resistance device that varies with temperature like the RTD?

8. What are the main advantages of the Kelvin bridge over the Wheatstone bridge in taking small resistance readings such as in fault location?

9. How does the thermocouple test set shown in this chapter read out temperature from units under test when in the millivolt test position?

10. How does the logic probe indicate whether a circuit is either high or low?

11. What method does the logic pulser use to inject one pulse into a circuit?

Troubleshooting and Testing

Using a Wheatstone bridge and a test procedure called the Murray Loop, it is possible to accurately measure the distance from the test instrument to the point in the conductor where a fault occurs.

A test technician is assigned such a task when one of the conductors in a long run of an existing lead-covered aerial control cable develops a ground. Construction at different points of the cable makes it difficult to pinpoint where the damage might have occurred.

From a terminal point at the end of the conductor and using the diagram shown in Figure 5–10, the faulty conductor is identified. Next, as shown in Figure 5–11 a known, good conductor is tied to the faulted conductor in a terminal box at the far end of the cable.

With the RVM switch set to M (Murray), resistance readings are taken through the faulted conductor and back through the ground to the test set. Multiplier dials are set at M1000 (A in the formula), and a reading of 76 ohms is obtained (R in the formula).

Next the RVM switch is set to R (resistance), and multiply by dial to 1/1. Readings through the good conductor and returning on the faulted conductor show a total reading of 283 (r in the formula).

QUESTIONS:
1. Using the formula below, solve for the ohms from the instrument to the fault (Xa in the formulas).

$$Xa = \frac{R \times r}{A + R}$$

2. Using Table 5–1 (page 118), solve for the distance in feet if the ohms in Xa are taken with number 22 wire at 68°F.

ANSWERS:
1. $\dfrac{76 \times 283}{1000 + 76} = \dfrac{21280}{1076} = 19.80$ Ohms
2. From Table 5–1, #22 wire = 61.95 feet per ohm
   Fault = 61.95 x 19.8 = 1226.6 feet from test instrument

Oscilloscopes 6

INTRODUCTION

The oscilloscope is a general purpose, cathode-ray tube (CRT) instrument that is useful in many applications. It provides a means of displaying how any voltage or current varies over time, by using a beam of electrons that strike a fluorescent screen to display a waveform. Any signal that can be converted to a voltage level can be displayed. When viewing a sine wave, it displays instantaneous values that occur during the full cycle, by showing a continuous display of one or more cycles. The unit shown in Figure 6–1 is representative of a typical single-trace oscilloscope.

OBJECTIVES

After studying this chapter, the student should be able to:

- *Understand how the oscilloscope operates.*
- *Know how to use the various oscilloscope controls.*
- *Know how to interpret the patterns displayed on the screen.*

FIGURE 6–1
Single-trace oscilloscope. (Courtesy of EMCO Division, Components Specialties, Inc., Lindenhurst, New York 11757)

PRINCIPLE OF OSCILLOSCOPE OPERATION

The oscilloscope is used to observe waveforms, measure frequency, and measure voltage. It is useful in many applications, including output indication, precise measurement of electrical pulses, and troubleshooting. Loading of the circuit measured is practically negligible. It is used to make amplitude and time measurements of any portion of a signal. Signals can vary from DC to a frequency of several million cycles per second (Hz). The range of frequencies a particular model can measure is called its bandwidth. Figure 6–2 shows the circuit of a typical oscilloscope. Inside the cathode ray tube (CRT) an electron gun aims its beam towards the screen. On the inside of the tube a phosphor coating glows when struck by the electrons.

Before the electrons can get from the gun to the screen, they must make their way through electrostatic fields set up by two sets of high-voltage plates, as shown in Figure 6–3.

FIGURE 6–2
Oscilloscope circuit.
(Courtesy of EMCO
Division,
Components
Specialties, Inc.,
Lindenhurst, New
York 11757)

Height of the display (Y axis) is controlled by the vertical plates in response to strength of the input signal. Horizontal plates control the width (X axis). This value equals the time span of the measurement, as selected by the operator.

By changing the polarity and amplitude of the voltages to both sets of plates, the shape and location of the display can be controlled. Using the graticule lines on the front of the CRT, as shown in Figure 6–4, accurate measurements are easily made.

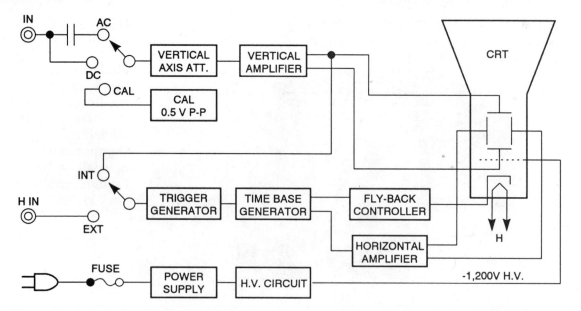

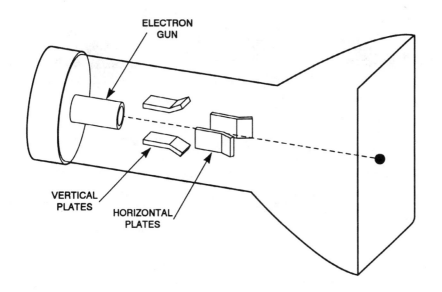

FIGURE 6–3
CRT tube

OPERATING THE SCOPE

Before connecting the power lead of an oscilloscope, make sure the POWER switch is in the OFF position and the intensity control is fully counterclockwise. Connect to a source of power, turn on, and allow a warm-up period of one to five minutes before making any adjustments.

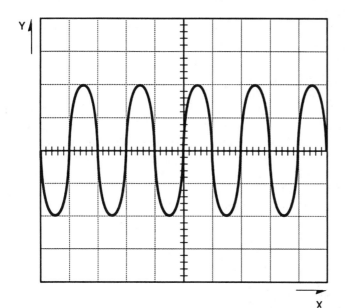

FIGURE 6–4
Oscilloscope screen

FIGURE 6–5
Oscilloscope probes

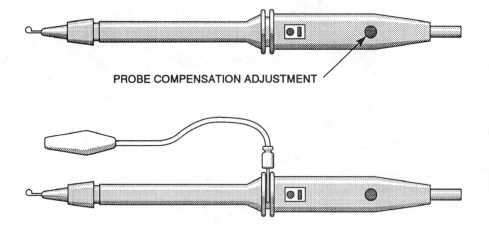

PROBE COMPENSATION ADJUSTMENT

Using an oscilloscope to measure voltage between two points is similar to using a multimeter. Oscilloscope probes use a coaxial cable. The input tip is connected through the body of the probe to the input on the oscilloscope, which is considered the positive side. The other lead connected to the metal sheath of the coaxial cable is the negative lead. This is usually connected to the ground side of the 120-volt AC supply that powers the oscilloscope. Care must be taken, when connecting the probe, to avoid erroneous readings and possible damage to the oscilloscope or the probe.

Two common types of oscilloscope voltage probes are used, Figure 6–5. The most common incorporates a voltage-dividing circuit (attenuator), either in the body of the probe or at the end of the cable. This reduces the input voltage and extends the maximum voltage that can be fed to the oscilloscope. Probes that allow one-tenth the value of voltage at the probe tip to reach the input, called X10 probes, are the most common, although X100 probes and others are available. This multiplying factor must be taken into account when reading the oscilloscope. Another common voltage probe is the direct probe, which does not have an attenuating circuit and is used for lower voltages. These low voltages utilize the maximum sensitivity of the oscilloscope. Direct probes are usually marked X1.

OSCILLOSCOPE CONTROLS

Figure 6–6 shows typical controls found on most oscilloscopes. Their identity and use are as follows.

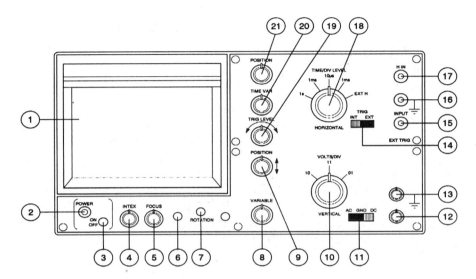

FIGURE 6–6 Oscilloscope controls. (Courtesy of EMCO Division, Components Specialties, Inc., Lindenhurst, New York 11757)

1. CRT DISPLAY. A five-inch CRT display is divided into eight divisions vertical, and ten divisions horizontal. Every blank is a 1 cm^2 square, with five equal divisions in each square.
2. POWER SWITCH. Pushed to turn power on. Pushed again to turn power off.
3. POWER INDICATION LAMP. Power is applied to the scope. The instrument is operating, and the power is in the "on" position.
4. INTENSITY. Controls the brightness of trace in CRT display. Brighter if adjusted clockwise; darker counterclockwise.
5. FOCUS. Controls the focusing of the trace, usually adjusted for sharpest image.
6. TRACE ROTATION. Adjusts the trace line deviation, induced by the earth's magnetism, until it aligns with the horizontal graticule lines.
7. CALIBRATOR. The calibrator terminal is the output of a high impedance square wave generator that is normally used to check the frequency response of the scope probes. The signal is usually about 1,000 hertz and about one half-volt peak-to-peak. When connected to the calibrator, probes that have a frequency response adjustment can be adjusted so that the waveform is very nearly a perfect square wave, like that shown in Figure 6–7A on page 140.
8. VARIABLE. Adjusts the magnitude of input so as to be more easily observed. Usually used with VERTICAL INPUT KNOB (10) to

obtain the precise magnitude. Adjust clockwise to maximum, then calculate the voltage according to the scale of VERTICAL INPUT KNOB (10).

9. POSITION. Controls the vertical position of the trace.

10. VERTICAL INPUT KNOB. A volts/Div. Knob in a 0.1, 1.0, 10.0 sequence to indicate the deflection factors. Set this knob in a higher range when the magnitude of input waveform is high, and in a lower range if the magnitude is low. The input voltage should be more than 10 mVp-p, and less then 600 Vp-p.

11. VERTICAL INPUT COUPLING DC-GND-AC selector. In the DC position, all AC and DC signals will pass to the input amplifier. In GND position, the input signal is not allowed to pass to the input. This position is used to set the zero reference of the scope trace. In AC position, only the AC signals will be coupled into the input.

12. VERTICAL INPUT CONNECTOR. A BNC connector with 1 MΩ input impedance, for application of external signals to the vertical input. This input is suitable for BNC test leads (including BNC test probe). The maximum input voltage is 600 Vp (AC + DC).

13. GROUND. The ground of the cabinet can be connected by a test lead or to the earth ground.

14. TRIGGER MODE (INT, EXT). In "INT" position, the trigger signal is generated by the internal circuit of the instrument. In "EXT" position, the trigger signal should be applied by external circuits. The external triggering is for synchronizing of complicated or irregular signals and should be used with TRIGGER LEVEL (19).

15. EXT TRIGGER INPUT. Input connector for external trigger signals. Suitable for banana-clip test lead.

16. Terminal ground for vertical or horizontal inputs. Suitable for banana-clip test lead.

17. HORIZONTAL. Input for horizontal signals; the input sensitivity is 200 mv/Div with 100 Vp maximum input voltage. Suitable for banana-clip test lead. Should be used with HORIZONTAL SWEEP RATE (18) set in EXT H range and H GAIN (20).

18. HORIZONTAL SWEEP RATE. A TIME/DIV switch in 1 µs-10 µs-.1 ms sequences to select the sweep rate should be used with (18) VAR. Set this switch in ms range to reduce the width of the waveform or to enlarge the width to the µs range. In the EXT H position the trace will not sweep.

19. TRIGGER LEVEL. Sets where on the slope the set will trigger. Adjusted clockwise will set it positive-going. The more it is adjusted in the right-hand direction, the higher on the slope the triggering will occur. Adjusted counterclockwise will set it negative-going. The more it is adjusted in the left-hand direction,

the higher on the slope the triggering level occurs. This control is
used to stabilize small signals.
20. H GAIN. A fine-tune control to adjust the length of trace, or to
control the external horizontal input signals. Adjusting clockwise
to the CAL position (maximum) to comply with HORIZONTAL
SWEEP RATE (18) will allow for precise time and frequency testing. The value can be calculated according to the position of HOR-
IZONTAL SWEEP RATE (18) and the scale of the screen.
21. ↔ Position, to adjust the horizontal location of trace.

OPERATING THE TEST SET

Line Voltage

- This instrument permits ±10 percent tolerance of the specified
operating voltage. The instrument will work improperly or may
be damaged if not operated in this range.
- Before powering on this instrument, please be sure to use the
correct line power and fuse.
- To prevent external interference and high voltage shock, the cabinet must be grounded.

Trigger

Internal sensitivity	0.4 DIV 30 Hz to 2 MHz
	1.5 DIV 2 MHz to 10 MHz

Maximum Input Voltage

Vertical input	600 Vp (DC + AC)
Horizontal input	100 Vp (DC + AC)
EXT TRIG input	30 Vp (DC + AC)
Z-Axis input	10 Vp (DC + AC)

Ion Burn on the Screen

Keeping the electron beam of the CRT in a fixed point on the screen
(such as happens when the HORIZONTAL SWEEP RATE [18] switch is
set in EXT H position) will cause an ion-burn on the screen. To prevent
this from happening, you should keep the trace moving or adjust the
INTENSITY CONTROL (4) fully counterclockwise to reduce the
brightness.

Earth Magnetism

Different geographical locations will cause a tilt angle to the trace line with regard to horizontal graticule lines. For precise adjustment, the inputs should be grounded, and set switch (11) to GND.

> **NOTE:** Do not operate the instrument under high-density magnetic fields, as this causes distorted waveforms.

PREPARATIONS FOR OPERATIONS

Turn on the power, and refer to Figure 6–6 on page 135.

Setting Before Operation

1. Set the vertical input (11) to GND.
2. Set the trigger mode (14) to INT.
3. Set the horizontal sweep (18) to 10 mS.
4. Set the intensity (4) clockwise to two o'clock position.
5. Adjust the position control (9), (21) to align the trace in the proper position on the screen.
6. Adjust the intensity control (4) and focus control (5) until the trace is satisfactorily clear.
7. Adjust the trace rotation (6) to align the trace with the horizontal graticule line of the screen.

Using Internal Trigger Signal

1. Set the trigger mode switch (14) to INT position.
2. Connect the test lead to vertical input connector (12); apply the external signal.
3. When testing AC signals, set the vertical input coupling switch (11) to AC position. When testing signals with AC, DC components or pure DC signals, set the Vertical Input coupling switch to DC position.
4. If the waveform displayed on the screen is not the proper magnitude, adjust the vertical input knob (10) and variable (8) to obtain a proper waveform.
5. If the waveform displayed in the screen is not the proper width, adjust the horizontal sweep rate control (18) and H gain (20) to obtain a proper waveform.

6. If the waveform cannot be properly synchronized, adjust the trigger level (19) until the proper waveform is displayed.

Using External Trigger Signal

An external trigger signal must be used when an irregular waveform cannot be synchronized by internal trigger signal.

1. Set the trigger mode switch (14) to EXT position.
2. Connect the test lead to the vertical input connector (12), and apply the external signal.
3. Connect the shunt circuit of the external signal source between the EXT TRIGGER INPUT (15) and GND (16).
4. Repeat steps 3–6, as described under Using Internal Trigger Signal.

Using the Calibrator

The output of the calibrator (7) is a 1,000 Hz square wave with 0.5 Vp-p voltage. Accuracy of the scale can be checked by setting the switches of the instrument to the proper positions, as shown in the following procedures:

1. Set the TRIGGER MODE switch (14) in INT position.
2. Connect one side of the test lead or the test probe (set in X1 range) to the VERTICAL INPUT connector (12).
3. Connect the other side of the test lead, test probe to the CALIBRATOR (7) output.
4. Set the VERTICAL INPUT COUPLING switch (11) to DC position.
5. Set the VERTICAL INPUT knob (10) to .1 V/Div position; then tune the VARIABLE (8) to CAL (maximum).
6. Set the HORIZONTAL SWEEP RATE switch (18) to 1 mS/Div position; then tune the H GAIN (20) CAL to maximum.
7. Now observe the voltage value displayed in the horizontal axis. There should be one period of square wave in every division.

   NOTE: A vertical axis calibration will be done to calculate the voltage value. A horizontal calibration will be done to calculate the frequency or period. The formula $F = 1/T$ can be utilized to calculate the frequency of the waveform.

Specifications for Suitable Test Probes

The frequency response for the vertical input of the test set is from DC to 10 MHz; therefore a suitable test probe must at least meet this specification. The maximum tolerable voltage is 250 Vrms or 600 VDC (in X10 position). Input impedance is 10 MΩ and 1 MΩ to the oscilloscope, and internal capacitance is 20–40 pf adjustable. When in X10 position, frequency response for ±2 percent accuracy is DC-15 MHz and up. In X1 position, the input impedance is 1MΩ (similar to the input impedance of the instrument), and internal capacitance within 250 pf frequency response is DC-5 MHz and up.

Calibrating the Test Probe

Connect a test probe of adequate frequency range, and the test set in X10 position. The output of the calibrator is a pure square wave signal. However, due to capacitance inside the test probe, the output waveform may be distorted, and the scope display may be similar to Figure 6–7, A through D.

To compensate for the high frequency loss and peaking, as shown in Figure 6–7B and 6–7C, a variable 20–40 pf capacitor is included in the

FIGURE 6–7
Comparison between correct and incorrect waveforms. (Courtesy of EMCO Division, Components Specialties, Inc., Lindenhurst, New York 11757)

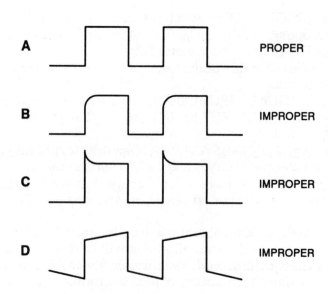

X10 range of the test probe. By adjusting the capacitor, the correct waveform, as shown in Figure 6–7A, will be obtained.

The screen will display incorrect waveform, shown as Figure 6–7D, if the VERTICAL INPUT COUPLING switch (11) is not set in DC (direct coupling) position. The deficient low-frequency condition is induced because the capacitor of AC position blocks the DC component.

NOTE:

1. A test probe with bad isolation will induce noise interference.
2. A test probe with bad frequency response will seriously attenuate high-frequency signals (more attenuation for higher frequency).
3. The test probe should be set in X10 range when operating in areas of strong background interference.
4. When the probe is set in X10 range, the magnitude displayed on the screen is one-tenth of its real magnitude.
5. An overly long test lead will induce high-frequency attenuation.

Using the Trigger Level Controller

The test set is designed with an internal automatic triggering circuit that will automatically control the trigger level for high-voltage vertical input. For a critical low-voltage input, this must be done by manual adjustment. Adjusting clockwise from the central "0" position will increase the positive-going trigger level. When adjusted counterclockwise, the negative-going trigger level will increase.

GENERAL APPLICATION EXAMPLES

Few test instruments are as versatile as the oscilloscope in measuring electrical and electronic values. Below are some examples.

Pure AC Signals Measurements

To measure the equalized continuous waves, such as sinusoidal waves, square waves, or triangle waves, internal triggering may be utilized.

AC Voltage Measurements

Example:

VOLTAGE INPUT KNOB set in 0.1 V/Div; VARIABLE set in CAL (maximum). The waveform displayed on the CRT screen is shown in Figure 6–8. Voltage of the waveform is calculated as shown below:

Real input voltage Vin = $0.1 \frac{V}{Div} \times 6 = 0.6$ V (Peak-to-peak)

Converted to Root Mean Square Values

$$VRMS = \frac{\text{Peak to peak V}}{2 \times \sqrt{2}} = \frac{0.6}{2 \times 1.414} = 0.212V \text{ (RMS)}$$

AC Frequency Measurement

Example:

HORIZONTAL INPUT switch set in 10 µs position, VARIABLE set in CAL (maximum) position; the frequency of waveform shown in Figure 6–8 is calculated as follows:

FIGURE 6–8
AC voltage and frequency measurement. (Courtesy of EMCO Division, Components Specialties, Inc., Lindenhurst, New York 11757)

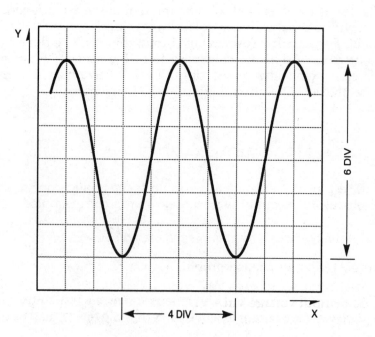

Real period of input waveform Tin = 10 μs/Div × 4 Div = 40 μs (1 cycle)

Utilizing formula f = 1/T, transfer T in to frequency

Real frequency of input waveform F in = 1/T in $\dfrac{1}{40\mu s}$ = 25KHz

DC Components Measurements

To measure equalized continuous waves containing DC components, internal triggering can be used. However, the initial horizontal position of the trace should be set and releaded in advance. To set the initial position, set the VERTICAL INPUT COUPLING switch (11) in GND position. Then tune the POSITION controller (9) to control the position of trace. If the input voltage is anticipated to be high, the VERTICAL INPUT KNOB should be set in the proper position. Also be sure to set the VARIABLE switch in CAL position.

AC and DC Components Measurements

When the test probe is set in X10 range, the measured circuit voltage shown in the screen of the instrument will be only one-tenth of the real value.

Example:
(1) Set the utilized test probe in X10 range.
(2) Input the DC signal containing the AC component.

VERTICAL INPUT KNOB set in 10 V/Div; VARIABLE tuned in CAL position. The waveform displayed on the screen of the instrument is shown in Figure 6–9.

Level including AC
Level for pure DC
Initial
Initial trace position

The voltage of the waveform is calculated below:

Real input DC voltage VDC = 10 V/Div × 4 Div = 40V
Real input AC voltage VAC = 10 V/Div × 1 Div = 10V (Peak)
Total input voltage Vin = VDC + VAC = 40V + 10V = 50V (Peak)
Voltage of the circuit Vk = Vin × 10 = 50V × 10 = 500V (Peak)

FIGURE 6–9
AC and DC
waveform. (Courtesy
of EMCO Division,
Components
Specialties, Inc.,
Lindenhurst, New
York 11757)

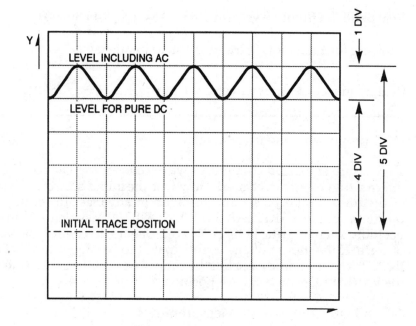

NOTES: (1) The measured voltage level displayed above the initial trace position is a positive voltage. Level displayed below the initial trace position is a negative voltage. The voltage shown in Figure 6–9 is a positive voltage.

(2) For pure DC signal, the AC pattern will not appear. The display is similar to the trace and will be above or below the initial trace position. The calculation method is the same as described above.

Irregular Signals Measurements

To measure signals that are not equalized continuous waves, such as pulses, TB synchronizing signals, or integration waves, the internal trigger synchronization cannot be applied. External triggering should be utilized to stabilize the displayed waveform.

Obtaining External Trigger Signals

A sophisticated signal generator is usually equipped with a synchronous signal output. This can be utilized as a trigger signal source for the oscilloscope. In general, the trigger signal from a voltage divider circuit can also be used. It should be noted that a higher trigger signal voltage results in better synchronizing performance. However, input voltage should not be exceeded.

V_s is the voltage source to be measured and is connected to the VERTICAL INPUT of the instrument. V_0 is obtained from R_2 of a voltage divider and is connected to the EXT TRIGGER INPUT of the instrument as a trigger signal. The magnitude of the divided voltage is decided by the ratio between R_1 and R_2.

Pulse Duration Calculation

HORIZONTAL SWEEP RATE switch is set in 10 μs position; VARIABLE is tuned to CAL position. The pulse displayed on the screen is shown in Figure 6–10. Pulse duration can be calculated as follows:

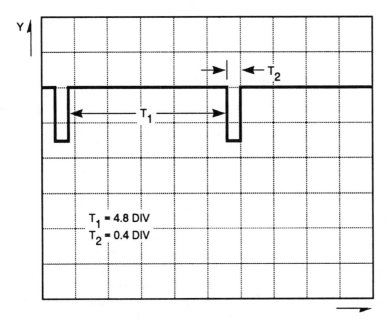

$T_1 = 4.8$ DIV
$T_2 = 0.4$ DIV

FIGURE 6–10
Horizontal sweep rate. (Courtesy of EMCO Division, Components Specialties, Inc., Lindenhurst, New York 11757)

FIGURE 6–11
Measuring current.
(Courtesy of EMCO
Division,
Components
Specialties, Inc.,
Lindenhurst, New
York 11757)

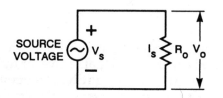

R_o = A RESISTOR OF FIXED VALUE
V_s = SIGNAL SOURCE THROUGH R_o

T_1 = 4.8 Div
T_2 = 0.4 Div

Duration of T_1 = 10 μs/Div × 4.8 Div = 48μs
Duration of T_2 = 10 μs/Div × 0.4 Div = 4 μs

Current Measurements and Calculation

Although an oscilloscope is usually used to measure voltages, Figure 6–11 shows a simple way to obtain current values.

Source Voltage R_0 = A resistor of fixed value
                        V_S = Signal source through R_0

A signal V_0 is picked and connected to the vertical input of the instrument to obtain the voltage; then from formula $I_S = V_0/R_0$, the current I_S can be calculated.

Amplitude Modulation Percentage Measurement

Please refer to the section on Settings Before Operation for control positions. Connect a carrier signal from a signal generator or from a transmitter to the VERTICAL INPUT. Proper voltage should be above 100 mVrms, and carrier frequency below 10 MHz. Figure 6–12 shows the formula to calculate amplitude modulation percentage.

DUAL TRACE OSCILLOSCOPES

The oscilloscope described above can be used to make many simple measurements required of electrical workers and instrument technicians. However, when amplitude phase or time comparisons need to be made, multiple trace scopes are required. Logic analyzers are multi-trace scopes that can display 16, 32, and even more traces of digital logic information simultaneously. The most common and versatile scope in use today is the dual trace oscilloscope, Figure 6–13. Most of the con-

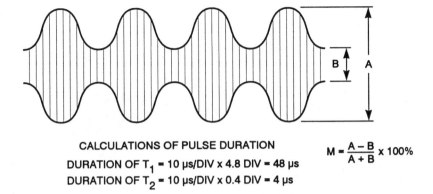

CALCULATIONS OF PULSE DURATION
DURATION OF T_1 = 10 µs/DIV x 4.8 DIV = 48 µs
DURATION OF T_2 = 10 µs/DIV x 0.4 DIV = 4 µs

$$M = \frac{A - B}{A + B} \times 100\%$$

**FIGURE 6–12
Amplitude
modulation
percentage
measurement.
(Courtesy of EMCO
Division,
Components
Specialties, Inc.,
Lindenhurst, New
York 11757)**

trols found on dual trace scopes have already been described. However, under the section on the single trace oscilloscope, there are three basic differences between the two.

1. There are two selector controls, one for Channel A, the second for Channel B—provided for amplitude, variable/calibrate, and input selection control.
2. A channel mode selector switch controls the presentation of the traces on the CRT. Available choices might include Channel A, Channel B, Alternate, Chopped, and A+B. The first two are used to select either Channel A or Channel B. These modes are desirable when attention must be focused on one signal at a time. Alternate means that Channel A and Channel B are displayed sequentially, first one then the other. This mode must be used at high frequencies when the chopping signal interferes with the signal being viewed. Chopped means that samples A and B are displayed alternately during each trace. The signals are chopped and displayed at high frequency so that they appear to be on the screen simultaneously. The chopped mode is very useful when the signals on both channels are below the chopping frequency. Another benefit of the chopped mode is that the phase relationship between the signals on the two channels is usually very reliable. The A+B mode displays the algebraic sum of the two channels. In this mode, if Channel B is inverted, the display shows the algebraic difference between the two signals.
3. The final difference is that there is an added function on the trigger selector. Instead of just internal and external trigger positions, there are positions to select Channel A, Channel B, or external as the source of the trigger.

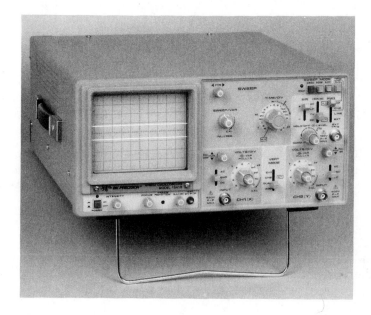

**FIGURE 6–13
Dual trace
oscilloscope.
(Courtesy of B+K
Precision)**

SUMMARY

Devices used in industry to measure chemical processes or power flow are becoming increasingly complicated. More and more of these measuring circuits are mounted on printed circuit boards and use integrated circuits. An oscilloscope is an excellent tool for following the chain of measurement values as an aid in troubleshooting equipment.

An oscilloscope has the ability of taking voltage and displaying its intensity and character over time. By measuring and calibrating a known signal, the value of an unknown signal can be measured and studied. Peak-to-peak voltage can be displayed, and any distortion of the signal is easily recognized. Any measurable quantity that can be converted to an electrical impulse can be measured with the oscilloscope.

REVIEW QUESTIONS

1. What does the Y axis represent on an oscilloscope?

2. What does the X axis represent?

3. Is there a simple way for the oscilloscope to read current?

4. If you are using a X10 probe, do you multiply or divide your displayed results by 10?

5. What does the trigger mode level control do?

6. How does the slope control work with the trigger control level control?

7. Is an AC sine wave displayed as peak-to-peak, rms, or average voltage?

8. How do you come up with rms voltage using an oscilloscope?

9. What is the term that describes the range of frequencies a particular model oscilloscope will handle?

10. Name an example of how a dual trace scope could be used.

Troubleshooting and Testing

Number 1

Troubleshooting a piece of electronic equipment usually begins with a good visual check, followed by a check of power and fuses. Checking power should include measuring the DC level and the AC ripple voltage at the output of each of the unit's power supplies to ensure that all are within tolerance.

Object: Measure the DC level and the AC ripple of a 5-volt supply.

When the scope is on and stabilized, obtain either a direct (x1) or a x10 probe and connect it to the Channel A amplifier input. Connect the probe tip to the scope's calibrator. Adjust the horizontal sweep rate to 0.5 millisec/div. and the Channel A amplitude to 0.1 volt/div. for a x1 probe, or to 0.01 volt/div. for a X10 probe. Adjust the trigger controls to obtain a stable waveform on the scope display. (Source to Channel A and Mode to Auto if the scope is so equipped.) Adjust the focus and intensity controls so that a clear, sharp trace is observed. If the probe has a calibration adjustment, use it to obtain a square wave that is free from any droop or peaking on the leading edge. Finally, change the Channel A amplitude to 1 volt/div. for a X1 probe or to 0.1 volt/div. for a X10 probe, change the Channel A input switch to 'gnd,' and then adjust the Channel A vertical position control so that the trace is on the bottom line on the face of the scope. As shown in Figure A, this is now the line that represents zero volts. Change the input switch to DC, and the scope is ready to make the first measurement.

Checking the DC level, remove the probe from the calibrator output and connect it to any point where the 5 volts are present. Starting at the zero line, count the number of divisions and tenths of divisions between the zero line and the scope trace. If the AC ripple is a significant measure from the zero line to the midpoint of the AC ripple, the DC voltage is:

DC voltage = # of divisions x 1 volt/div. (X1 probe)

   or

DC voltage = # of divisions x 0.1 volt/div x 10 (X10 probe)

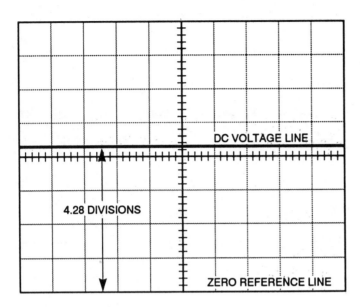

4.28 DIVISIONS

DC VOLTAGE LINE

ZERO REFERENCE LINE

FIGURE A
(*Courtesy of Ploehn Engineering and Consulting*)

Checking the AC ripple, leave the probe where it is. Change the channel input switch to 'gnd' and adjust the vertical position control so the trace is on the center line of the display; this is the new zero line as shown in Figure B. Change the input switch to the AC position and change the channel amplitude control to a lower volts/div. position so that an accurate measurement of the number of divisions can be made. The AC component or ripple can be found as follows:

AC ripple = # of divisions x # volt/div. (X1 probe)

or

AC ripple = # of divisions x # volt/div. x 10 (X10 probe)

NOTE: The amplitude of volt/div. control on some scopes is marked for both X1 and X10 probes so that the X10 factor used in the calculations above is not necessary.

FIGURE B
*(Courtesy of Ploehn
Engineering and
Consulting)*

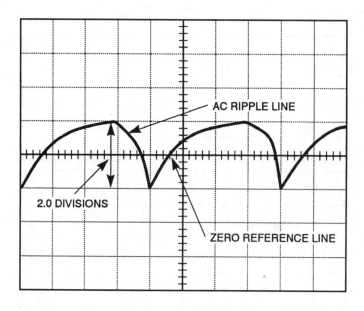

Number 2
    Design verification and troubleshooting often require the tech-
nician to compare the phase and amplitude of amplifier inputs
and outputs. This type of measurement is easily done with a dual
trace oscilloscope.

    *Object: Measurement must be made of the time delay between the
leading edge of a trigger pulse and the trailing edge of the output pulse
from a flip-flop circuit. The pulses are in the TTL logic circuits operating
at 5 volts. Both pulses go from 1 to +4.5 volts and should be separated by
about 20 microseconds.*

    Turn the scope on and, once stabilized, connect two X10 scope
probes, one to Channel A and the other to Channel B. One at a
time, check the calibration of each probe to ensure the frequency
response is correct. Connect Channel A probe to the trigger source
and Channel B to the flip-flop output terminal. Set both channel
amplifiers to 2 volt/div. and the channel mode switch to the
'chopped' position. The 2 volt/div. selection is chosen so that each
pulse will be just over two divisions high, and the 'chopped' mode
so the phasing between the pulses will be correct. Ensure that both

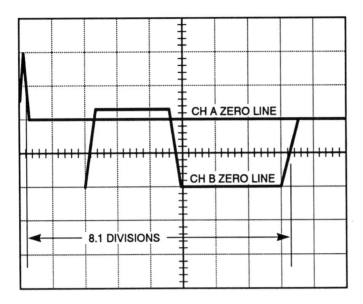

FIGURE C
(Courtesy of Ploehn Engineering and Consulting)

channel calibration switches are set to the 'Cal' position. Select the 'Auto' position of the trigger control. Set both input selector switches to the 'gnd' position, then adjust Channel A position to three divisions from the top, Channel B position to one division below the center line. Adjust the intensity and the focus controls for a sharp display, then set horizontal position control so the traces start at the left marking on the CRT. Set the time/div. control to 5 microseconds/div.

Making the Measurement

Turn the equipment power on and set the trigger source control on the scope to Channel A and the trigger slope control to the positive edge. If necessary, change the trigger mode control to 'manual,' then adjust the trigger level to produce a stable display. With the leading edge of the Channel A pulse on a vertical line, count the number of divisions and tenths to the trailing edge of the pulse on Channel B as shown in Figure C. The time measurement is found by multiplying the number of divisions measured by the time of the horizontal time/div. control.

NOTE: Time measurements like this are usually made at the 50 percent amplitude point of the pulse.

Watt, Watthour Phase-Angle Meters

7

INTRODUCTION

In the industrial world, with its tight budgets and bottom lines, it is important to measure power usage accurately. Not only does the electric utility need to know customer usage, but quite often factories have their own metering to monitor department usage, and kilowatt-hour meters are used as tools of measurement.

A phase-angle meter is a useful tool for determining the relationship between voltages and current in a circuit. One use of this instrument is to verify the connections of a watthour or kilowatt-hour meter installation. Another use is in checking the polarity of instrument transformers.

OBJECTIVES

After studying this chapter, the student should be able to:

- *Understand how to connect wattmeters to monitor circuits.*
- *Understand how theta affects AC circuits.*
- *Know how delta circuits operate.*
- *Know how wye circuits operate.*
- *Know how to calculate power factor using a kilowatt-hour meter.*
- *Understand the operation of the phase-angle meter.*

SINGLE-PHASE AC WATTS

You already learned that watts, the measure of work done, is equal to
E × I, when either DC or AC are put through a resistor. However, when
AC voltage is used to power a motor, or capacitors are added to the line,
voltage and current are no longer able to act exactly together.

In order to follow what happens in these types of electrical circuits,
engineers felt that analyzing a sine wave was too complicated. Instead,
they decided to use a circle to represent one cycle of AC voltage. If the
circle equals one cycle, then one time around the circle equals 1/60th of
a second.

When an AC motor is first connected to a voltage source, the coils, or
stator, are energized. This sets up a magnetic flux that is induced into
the rotor. After some time delay, the rotor begins to turn, and current
begins to flow in the circuit.

Figure 7–1 shows this time delay between the original voltage energizing the coils and the time when current starts to flow in the rotor. The
arrow shows the direction the voltage and current are spinning. You,
the observer, are at the three o'clock position, watching.

Remember that DC is a steady state voltage. Recall also that AC goes
through a cycle from zero to positive, to zero, then to negative, and
finally back to zero again. As a result, AC voltage does not have a steady
push on current (except in a resistor). In fact, as the voltage goes
through its cycle, sometimes the current cannot keep up with it. That is
what happens in the motor circuit described above.

FIGURE 7–1
Voltage and current
in an AC circuit

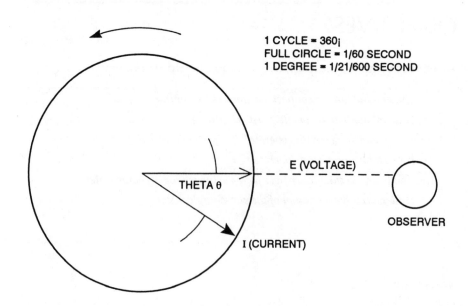

1 CYCLE = 360¡
FULL CIRCLE = 1/60 SECOND
1 DEGREE = 1/21/600 SECOND

E (VOLTAGE)

OBSERVER

THETA θ

I (CURRENT)

In a capacitive circuit, the opposite is true. Capacitive devices can operate so that current actually rushes ahead, or leads the voltage.

Theta (θ) is the term for this difference, in degrees, between the voltage and current. The fewer degrees of theta, the more efficiently the circuit is performing. The Greek letter *theta* (θ) may sound mysterious, but it simply represents a small measure of time measured in degrees of a circle.

How small a measure of time? Using your pocket calculator, find the decimal value of time for one cycle of AC voltage: that is, 1/60 of a second equals 0.01667 seconds. Next divide 0.01667 by 360 degrees. If the answer is displayed as 4.63-05, simply write down 4.63, and move the decimal point five places to the left:

0.0000463 seconds.

If you are interested in how long 35 degrees is in real time, for example, multiply $35 \times 0.0000463 = 0.00162$ seconds.

An easier way to work with theta is to use some simple trigonometric functions. You already know that the larger the angle of theta, the more time goes by between voltage pushing and current flowing. The more degrees (theta) between voltage and current, the less efficient the circuit, and the fewer watts produced. Since watts are actual work performed, any circuit that has zero degrees of theta, such as a resistive load, is the most effective.

If your pocket calculator has trigonometric functions, enter zero degrees (0). Next press the cos (cosine) key. Notice the number one (1) on the display. Multiply 1×100 to change the number to percent. The efficiency of a circuit with voltage and current exactly in phase (zero theta) equals 100 percent and is called a unity power factor load.

Suppose a typical motor circuit, as described earlier, had a theta of 30 degrees. Enter 30 on the calculator and press the cos key to read 0.866. Multiply that number times 100 and get a lagging power factor of 86.6 percent. Another word for the percent efficiency of an electrical circuit is POWER FACTOR. Power factor will always equal the cosine of theta.

To calculate power in any single-phase AC circuits, like those found in most homes or apartments, the formula is:

Watts = E × I × Cos theta

Example:
Volts = 120
Amps = 10

Theta = 30 Degrees
Watts = E × I × Cos theta = 120 × 10 × .866 = 1,039.2 Watts

BLONDEL'S THEOREM

In the last century, a French physicist named André Blondel came up with a formula to measure the power in any electrical circuit. It states:

1. Take the number of wires in the circuit to be measured (N) and subtract one (1) = N–1.

   Example: Two wires feeding a motor equals 2–1 = 1

2. This is the number of current coils needed, which must be in series with the line.

   Example: The current coil must be in series with one of the two wires feeding a motor.

3. The same number of potential coils as there are current coils must be installed. These coils must be fed from the wire where the current coil is installed and terminated to the wire where no current device is installed. Figure 7–2 shows a wattmeter installed on a two-wire circuit.

   Studying how a wattmeter is made shows that it measures power by following the power formula given above. Voltage is measured by a coil using many turns of fine wire in parallel with the circuit. The higher the voltage, the more flux is given off.

   Current is measured by a coil wound with heavy wire, and in series with the conductor carrying the load. As current rises and falls in the measured circuit, so does the magnetic flux in the current coil.

   The wattmeter now measures E with the potential coil, I with the current coil. As theta increases or decreases, E flux and I flux work more or less efficiently, to deflect the pointer up scale to read watts, the work performed.

SINGLE-PHASE WATTHOUR METERS

The purpose of this book is to learn about electrical test instruments. There might be some confusion as to why a single-phase electric meter, such as those used on most homes by the power company, would be included here. With an accuracy of 0.2 percent (read that as 2/10 of 1

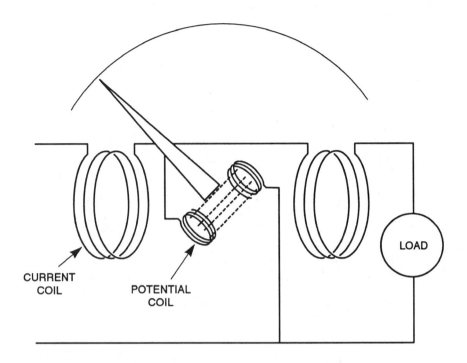

FIGURE 7–2
Wattmeter

percent), the kilowatt-hour meter is more accurate than most analog meters, and as accurate as most digital meters. Besides, you can use it to measure AC circuits for free.

Figure 7–3 shows a simplified drawing of the typical three-wire house meter. Notice the one potential coil and the two current coils. This meter does not follow Blondel's Theorem exactly. In order to make a less expensive meter, one potential coil has been eliminated.

As voltage and current flux are developed in the meter, they are induced in a meter disk, suspended between a bearing system that allows it to move with little friction. As load is drawn through the current coils, the disk moves within the poles of a permanent magnet. Strength of the magnet is calculated, so disk speed equals watts measured, and also introduces time, kilowatt-hours, into the measurement.

Motion of the disk is then transferred through gears to the register, where it is displayed on either cyclometer or dial types, to show kilowatt-hours used.

A cyclometer register is read just the way it appears, from left to right. Dial pointer-type registers are read following a slightly different procedure. Keep in mind that the numbers go up in value in multiples of ten from right to left. Figure 7–4 shows the register layout with units, the least significant figure being on the right. Also keep in mind that the

FIGURE 7–3
Meter element for
three-wire meter

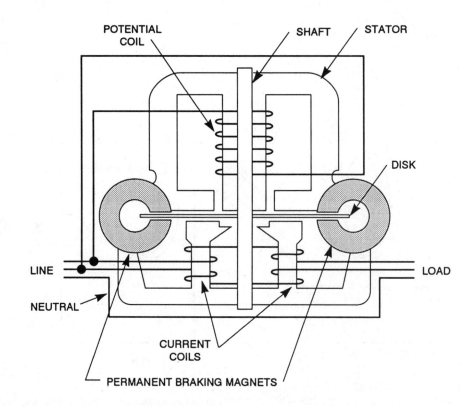

register is read, and the numbers written down, from right to left. Once the numbers are written down, they are read from left to right.

Until the first dial goes from 1–9 and finally passes 10, the next dial to the left cannot be read as the next whole number.

Although the reading might appear to be 80089, that is not correct.

FIGURE 7–4
Kilowatt-hour
register

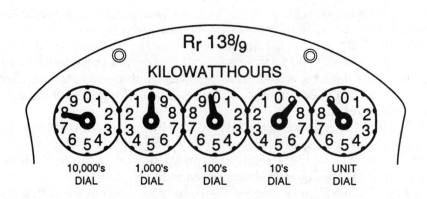

Read the register in this way:

1. The least significant number on the dial to the right is a 9.
2. This means that the next number to the left is not yet a 9; read it as 8.
3. Until that second dial goes past 0 the third dial cannot be 0, but is still 9.
4. The fourth dial also cannot be 0 before the third dial goes past 0, so it is still 9.
5. And, finally, the fifth dial cannot be 8 until the fourth dial goes past 0, so it reads 7.

The correct reading is: 79989.

In order to use a kilowatt-hour meter as a test instrument, certain nameplate values need to be understood. Figure 7–5 shows the nameplate of a typical three-wire house meter.

Cl 200 stands for 200-ampere class and is the load this meter can carry without overheating.

TA 30 is the industry-recommended test amperage (called the full load test) for calibrating this class of meter. In other words, testing the meter at 30 amperes will ensure meter accuracy from 200 amperes down to a few amperes. A second test carried out at 10 percent of full load (called the light load) ensures accuracy down to a few watts.

Kh 7.2 is called the meter test constant. It means every time the meter disk goes around one time, 7.2 watthours of power have gone through the meter. Knowing this, the kilowatt-hour meter can be used to measure any watt load within a home or apartment. All that is needed is a watch with a second hand, or a digital watch that displays seconds to time meter disk revolutions.

For example, to measure watts consumed by some appliance or motor, make certain that all other appliances are off, except the one to

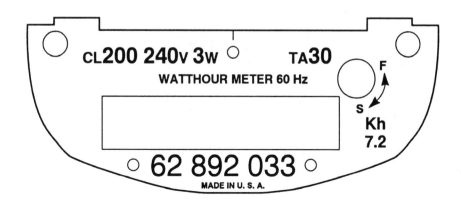

FIGURE 7–5
Meter nameplate

be measured. Watch the meter disk turn, and when the flag (dark spot) lines up with some point on the meter front (remember parallax), start keeping time. When the disk next comes around, that equals one revolution; but continue for a total of five or ten revolutions. Read the total elapsed time in seconds. Watts are then calculated using the following formula:

$$\text{Watts} = \frac{\text{Meter revolutions} \times \text{Meter kh} \times 3,600}{\text{Time in seconds for the above revolutions}}$$

Meter Kh is on the nameplate of the meter and can vary by age and model.

The number 3,600 is the number of seconds in an hour.

Example:

      Meter revolutions = 10
      Meter kh = 3.6
      Time in seconds = 100

$$\text{Watts} = \frac{10 \times 3.6 \times 3,600}{100} = \frac{129,600}{100} = 1,296 \text{ Watts}$$

Should the reading be required in kw, simply divide by 1,000 to equal 1.30 kw.

Suppose you needed to know the efficiency, or power factor, of the motor used in this appliance. Later in this chapter, the use of a phase-angle meter to obtain theta will be covered; but for now, the only instrument available is a clamp-on ammeter, with a built-in voltmeter.

While the appliance is running, carefully read and record the circuit voltage. Next, clamp the ammeter over one of the two wires supplying the appliance. Read and record the amperes.

Solving for power factor is based on the relationships within what is called the power triangle, shown in Figure 7–6.

To find the efficiency, or power factor, of a circuit:

$$\text{Power factor} = \text{Cosine of theta} = \frac{\text{Watts}}{\text{Volt amperes}} \times 100$$

From the example above:

      Watts = 1,296
      Volts = 119

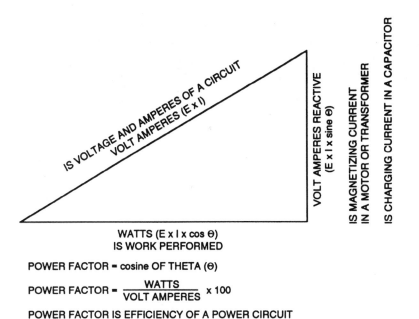

FIGURE 7–6
Power triangle

POWER FACTOR = cosine OF THETA (Θ)

POWER FACTOR = $\dfrac{\text{WATTS}}{\text{VOLT AMPERES}}$ × 100

POWER FACTOR IS EFFICIENCY OF A POWER CIRCUIT

Amps = 13
Volt amperes = 1,547 (119 × 13)

$$\text{Power factor} = \frac{\text{Watts}}{\text{Volt amperes}} = \frac{1{,}296}{1{,}547} \times 100 = 85 \text{ Percent power factor}$$

Vars, or volt amperes reactive in a circuit, is a measurement of the magnetizing current in a motor, or the electrostatic charge in a capacitive circuit. Sometimes called imaginary power, vars will not register on a wattmeter or kilowatt-hour meter. Only watts, real power, will register.

POLYPHASE AC CIRCUITS

Three-phase voltages can be generated by rotating three coils, evenly spaced, on a common rotor in a magnetic field, as shown in Figure 7–7A. As the coils rotate, voltages 120° apart will result, as shown in Figure 7–7B. How these voltage coils are tied together will determine if the system will be wye (Y) or delta (Δ).

Before trying to understand how these systems work it would be best to define the voltages and currents that make up three-phase circuits.

**FIGURE 7–7A
Coils of an AC
generator**

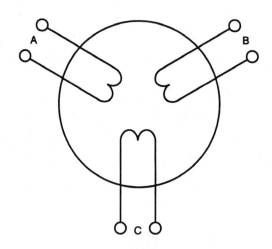

**FIGURE 7–7B
Three-phase sine
waves**

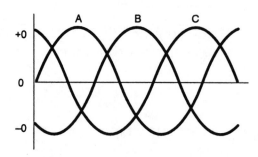

Definition	*Designation*
Phase voltage or phase current (Ep or Ip)	A to N
	B to N
	C to N
Line voltage or line current (El or Il)	A to B
(Also called phase-to-phase)	B to C
	C to A

Three-Phase, Three-Wire Delta

When the coils of a generator or transformer are delta-connected, they resemble the triangle shown in Figure 7–8. Currents flowing out of any conductor in this system are fed by two coils. Because the voltages are 120° apart, the currents do not double in value. Instead, current in the conductor will be 1.732 × the current in the coil. For a balanced system,

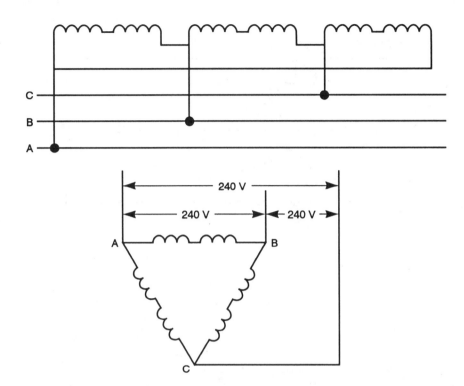

FIGURE 7–8
Three-phase,
three-wire delta

such as a three-phase motor, 10 amperes might flow in each coil of the transformer, but 17.32 amperes would be measured in the conductor leading to the motor.

According to definition, the only voltage available in a three-phase, three-wire system is phase voltage. At the equipment level this is usually 240 or 480 volts.

Should wattmeters be used to measure this three-wire circuit, according to Blondel's Theorem, two meters would be connected, as shown in Figure 7–9. Total power for the circuit would equal the readings of the two wattmeters added together.

Sometimes a condition exists where the power factor of a circuit feeding a number of motors or loads needs to be known. If all other circuits can be turned off, the following method can be used.

Because this circuit is three phase, some steps will be different from the single-phase circuit measured above. Once again, take a stopwatch check of the meter, with all load to be measured turned on. Calculate watts, just as was done in the single-phase circuit, using the formula:

$$\text{Watts} = \frac{\text{Meter revolutions} \times \text{Meter kh} \times 3{,}600}{\text{Time in seconds for the above revolutions}}$$

FIGURE 7–9
Wattmeters on a
Δ-connected load

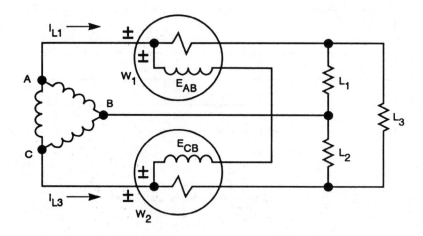

Next, take ammeter readings, and record currents in each of the three phases. Last, take readings, and record voltages from A to B, B to C, and C to A.

Now calculate the following:

>   Meter revolutions = 5
>   Kh = 28.6
>   Time =77 Seconds

$$\text{Watts} = \frac{5 \times 28.6 \times 3{,}600}{77 \text{ Seconds}} = \frac{514{,}800}{77} = 6{,}637 \text{ Watts, or } 66.37 \text{ Kw}$$

Calculate volt amperes by the following method.

Add all three voltages, and divide by three, to come up with an average voltage.

Phase	Voltage
A to B	235 Volts
B to C	237 Volts
C to A	<u>239 Volts</u>
	711/3 = 237 Volts

Add all three currents, and divide by three, to come up with average current.

Phase	Current
A	20 Amps
B	23 Amps
C	<u>20 Amps</u>
	63/3 = 21 Amps

Now apply the modified formula below for volt amperes in a three-phase, three-wire delta system.

Volt amperes = $1.732 \times E \times I = 1.732 \times 237 \times 21 = 8620$, or 86.2 kVA

$$\text{Power factor} = \frac{\text{Watts}}{\text{Volt amperes}} = \frac{6637}{8620} = .77 \times 100 =$$

77 Percent power factor

Three-Phase, Four-Wire Wye

Figure 7–10 shows the coil connections for a Y-system. One end of each coil is connected to a common point, and the other end to a transmission line conductor. Because each coil can operate by itself, current measured in the conductor will be the same as measured in the coil.

Notice that voltage can be read either phase-to-neutral or phase-to-phase in a Y system. At the equipment level, phase voltage is usually 120 volts, and line voltage is 208. However, other systems are used, such as 277/480 volts.

Using Blondel's Theorem, three wattmeters would be connected, as shown in Figure 7–11. Watts would equal the total readings from all three meters.

Newer electro-mechanical, three-phase, kilowatt-hour meters have two or three meter elements on one frame, with a common disk. Each element is equivalent to one wattmeter and has its own voltage and current coil.

Using the same method used for the three-phase, three-wire meter, a load check can be made using the power company meter, a voltmeter, and an ammeter.

Again, take a stopwatch check with the load to be measured turned on.

Take and record voltages between A to N, B to N, and C to N.

Take and record ammeter readings on all three phases.

**FIGURE 7–10
Three-phase,
four-wire wye**

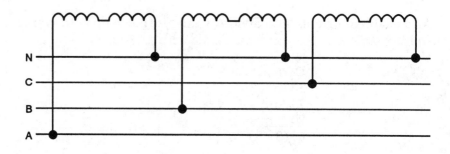

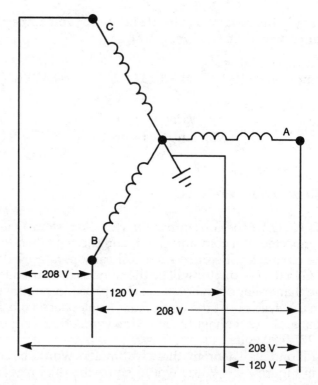

Example:

Meter revolutions = 10
Kh = 21.6
Time = 17 Seconds

$$\text{Watts} = \frac{10 \times 21.6 \times 3{,}600}{19} = \frac{777{,}600}{19} = 40{,}926 \text{ Watts or } 40.9 \text{ kw}$$

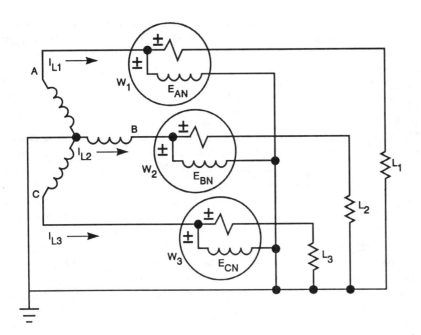

FIGURE 7–11
Wattmeters on a
Y-connected load

Add all three-phase voltages, and divide by three, for an average voltage.

Phase	Voltage
A to N	122 Volts
B to N	126 Volts
C to N	123 Volts
	371/3 = 123.7 Volts

Add all three currents, and divide by three, for average current.

Phase	Current
A	122 Amps
B	132 Amps
C	121 Amps
	375/3 = 125 Amps

Next, apply the formula for volt amperes in a three-phase, four-wire Y system:

Volt amperes = 3 × E × I = 3 × 123.7 × 125 = 46,388 or 46.4 kVA.

$$\text{Power Factor} = \frac{\text{Watts}}{\text{Volt amperes}} \times 100 = \frac{40,926}{46,388} = \times 100 = 88 \text{ Percent}$$

PHASE-ANGLE METERS

The phase-angle meter shown in Figure 7–12 can be used in a number of power system, shop, field, or laboratory applications. It will measure and display the phase angle relationship (theta) between two single-phase AC signals applied to its input circuit. Built-in batteries simplify measurements of 480-volt circuits.

Two identical input circuits are provided to facilitate measurements between two voltages, two currents, or a voltage and a current. Each circuit has binding posts for connecting the input signals. The two white binding posts, labeled COMMON ±, are used for both potential and current inputs. Although the two white binding posts are electrically the same point, the larger one should be used for currents in excess of 30 amperes. The smaller one can be used for currents up to 30 amperes. Either white binding post can be used as the common potential input. Current inputs are applied to the white common binding post and appropriate black binding post, as determined by the value of current. Potential inputs are applied to the white common binding post and appropriate red binding post, as determined by the value of potential.

The digital display indicates the phase angle in degrees (theta) that the signal applied to INPUT 1 lags the signal applied to INPUT 2. Once the

**FIGURE 7–12
Phase angle meter.
(Courtesy of AVO
Multi-Amp
Corporation)**

angle is known, it is a simple matter to translate this figure to actual time. A circle is equal to one sine wave, which is 1/60th (0.0167) of a second in duration. There are 360 degrees in a circle, so each degree equals 0.000046389 (0.0167/360) second.

PHASE-ANGLE METER CONTROLS

The following is a description of the controls and their operation.

Phase-Angle Meter

Power on switch—ON/OFF switch for input power for logic and display.

Zero check switch—Self-checks instrument calibration.

Hold reading switch—Instrument will retain whatever measurement is being displayed when this switch is turned on.

Display reset switch—When toggled, the battery operation is reset for approximately five (5) additional minutes.

Display test switch—All 8s are displayed to verify display operation.

Input 1 & 2 binding post—Display indicates, in degrees, the phase angle that input 1 lags input 2.

Hold Reading Feature

If it is desirable to retain the reading displayed, simply use the HOLD READING switch. Whatever measurement is being displayed at the time this circuit is activated will be held, even if input signal is removed, until the switch is turned OFF.

Zero Check Feature

Instrument calibration can be readily checked by use of the ZERO CHECK switch. To perform ZERO CHECK, allow instrument to warm up for 15 minutes. Then apply approximately 5 amperes to the 1.5–6.0 amperes range of INPUT 2 or approximately 120 volts to the 15–600 volts range of INPUT 2. No connection to INPUT 1 is required. Turn the ZERO CHECK switch ON, and the display should read between 359.8 and 0.2.

Phase-Angle Meter Applications

**FIGURE 7–13
Connections of a
three-phase,
four-wire Y,
three-element meter**

The phase-angle meter has many applications. One common use is verification of a kilowatt-hour or watthour meter installation. Quite often, a large, industrial complex will use meters to monitor electrical usage of various departments. When the service feeding these separate installations is over 200 amperes, or higher than 240 volts, current (CTs) and potential (VTs) instrument transformers are used. Rather than hard wiring directly to Class 10 or Class 20 transformer-rated meters, they often terminate at a test switch, as shown in Figure 7–13.

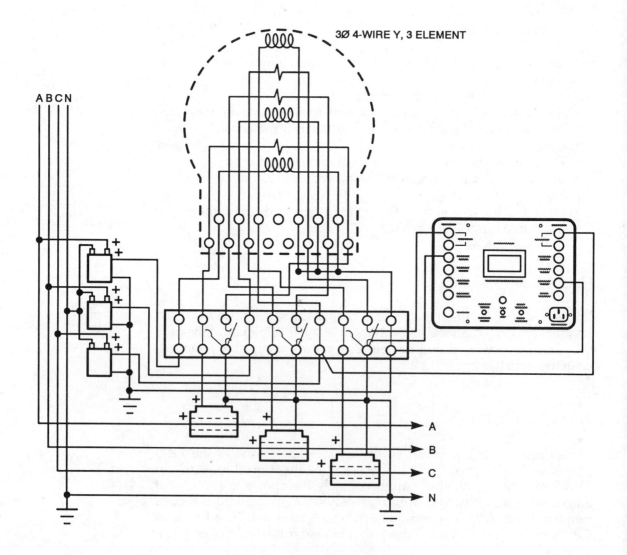

FIGURE 7–14
Test switch.
(Courtesy of AVO
Multi-Amp
Corporation)

Test switches come in many configurations and color combinations. On the switch shown in Figure 7–14, the current elements (dark handles), have an in position (+) on the left, and a test jack position on the return side, or right. Pulling down the left switch automatically shorts out the secondary of the CT for safety. Figure 7–15 shows schematically how this works.

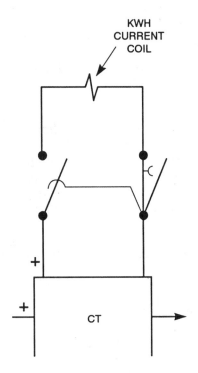

FIGURE 7–15
Test switch current
element

FIGURE 7-16
Dual circuit test plug

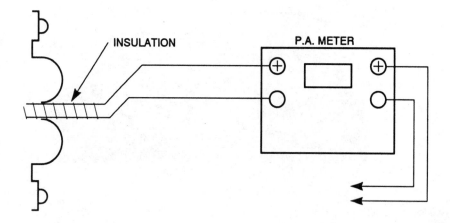

Theta is measured by connecting the current coil of the phase-angle meter in series with one meter current at a time, and reading its relationship to the potentials. Inserting a dual circuit plug into the test jack position puts the phase-angle meter in series with the current circuit. Figure 7–16 shows a side view of the test jack and how the dual circuit plug passes current to the phase-angle meter, and then returns it to the circuit.

For.safety, the following sequence of switch operations is suggested, when inserting the test plug:

1. Pull the left current-in switch down to short out the secondary of the CT.
2. Pull down the right test jack position switch.
3. Insert dual-circuit test plug, with leads connected to the phase-angle current coil, into the test jack position.
4. Close the left current-in test switch to obtain readings.

Use the following sequence to remove the test plug when moving to the next current:

1. Pull down the left current-in switch.
2. Remove the test plug.
3. Close the right test-jack switch.
4. Close the left current-in switch.

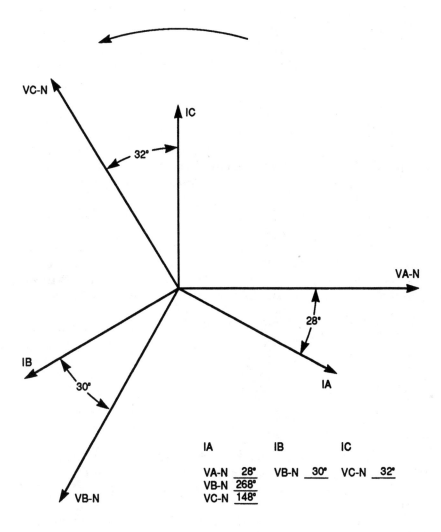

FIGURE 7–17
Three-phase,
four-wire wye work
sheet

	IA	IB	IC
	VA-N 28°	VB-N 30°	VC-N 32°
	VB-N 268°		
	VC-N 148°		

Figure 7–17 shows a suggested work sheet for taking phase-angle readings of a three-phase, four-wire, Y circuit. The dual-circuit plug is inserted in current A, and potential readings are taken in the sequence shown in the left-hand column. The lead connected to the plus side of the phase-angle potential coil always goes to the phase positions (three white handles to the left—A, B, or C); the return goes to the neutral (white handle on the right).

In similar manner, read current B and potential B–N. Last, read current C and potential C–N.

Since input 1 lags input 2, the three currents lag the potential by the degrees shown. Find the average theta as follows:

Phase	Theta
A	28
B	30
C	<u>32</u>
	90/3 = 30 Degrees

Power factor = Cosine of 30 degrees = 0.866 × 100 = 86.6 percent lagging power factor. This installation appears to be properly connected.

Suppose this was a capacitive circuit, with current leading the voltage. Since all readings show current, input 1, lagging the voltage, input 2, how would the meter display these results? One way to take such readings would be to switch current to input 2 and voltage to input 1, but that could be confusing.

Figure 7–18 shows how the phase angle affects a meter circuit. In a three-phase, four-wire, Y circuit, the kilowatt-hour meter will continue to spin and the wattmeter register, even if theta is 89°. As an example, measurements taken from a set of VTs and CTs show 120 volts at 5 amperes, or 600 volt amperes (120 × 5 = 600). The cosine of 89° is 0.017. Multiplying 0.017 × 600 = 10.2 watts results in few watts considering the high volt amperes, but still forward registration.

At 90°, the cosine is zero (0), and watts also equals zero. Both the kilowatt-hour and wattmeter will stop registering.

Both meters will try to register negative registration at 91°. At 180°, the kilowatt-hour meter will run full speed backwards, and the wattmeter will peg with fullest negative registration. The meters will start to slow negative registration at 181°.

When theta reaches 270°, the kilowatt-hour meter will stop spinning, and the wattmeter will indicate zero watts. If you have a pocket calculator with trig functions, you can check that the cosine of 270° does equal zero (0).

Once theta crosses 271°, both meters start to register watts, until at 360° the cosine is again 1, or 100 percent.

In order for the phase-angle meter to indicate a leading power factor, with current lagging voltage, it must display a reading between 270° and 360°.

In the real world, the only way theta between a current and its matching potential (IA and EA–N, for example) can be 91 to 269 degrees in a wye system is because of a wiring error.

DETERMINING CURRENT TRANSFORMER POLARITY

Using a current transformer of known polarity, a current transformer of unknown polarity can be tested using a phase-angle meter.

FIGURE 7–18
How theta affects
registration

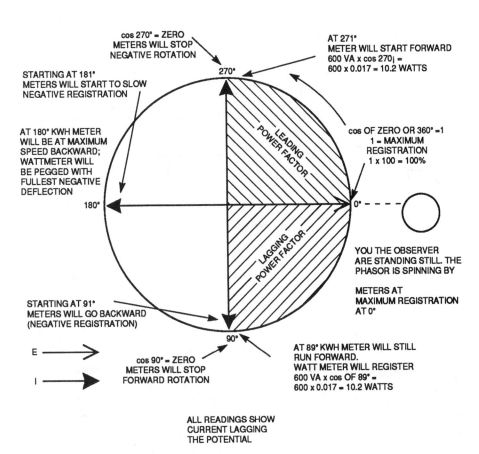

cos 270° = ZERO
METERS WILL STOP
NEGATIVE ROTATION

AT 271°
METER WILL START FORWARD
600 VA x cos 270¡ =
600 x 0.017 = 10.2 WATTS

STARTING AT 181°
METERS WILL START TO SLOW
NEGATIVE REGISTRATION

AT 180° KWH METER
WILL BE AT MAXIMUM
SPEED BACKWARD;
WATTMETER WILL
BE PEGGED WITH
FULLEST NEGATIVE
DEFLECTION

270°

LEADING
POWER FACTOR

cos OF ZERO OR 360° =1
1 = MAXIMUM
REGISTRATION
1 x 100 = 100%

180°

0°

LAGGING
POWER FACTOR

YOU THE OBSERVER
ARE STANDING STILL. THE
PHASOR IS SPINNING BY

METERS AT
MAXIMUM REGISTRATION
AT 0°

STARTING AT 91°
METERS WILL GO BACKWARD
(NEGATIVE REGISTRATION)

90°

E

I

cos 90° = ZERO
METERS WILL STOP
FORWARD ROTATION

AT 89° KWH METER WILL STILL
RUN FORWARD.
WATT METER WILL REGISTER
600 VA x cos OF 89° =
600 x 0.017 = 10.2 WATTS

ALL READINGS SHOW
CURRENT LAGGING
THE POTENTIAL

Connect the two current transformers to the phase-angle meter, as shown in Figure 7–19.

If the polarities are the same, a phase angle of 0° should be indicated. A reading of 180° shows that the polarity is opposite that of the known transformer.

SUMMARY

The wattmeter is used to measure true instantaneous power in an AC circuit. Single-phase wattmeters have two current terminals and two potential (voltage) terminals; one terminal of each pair is marked with a ± sign. Current from the source to the load passes in series with the wattmeter current coil. The voltmeter terminals are connected so the voltage coils are in parallel across the load.

> **WARNING**
> *Never open the secondary of a current transformer under load.*

FIGURE 7–19
Determining current
transformer polarity

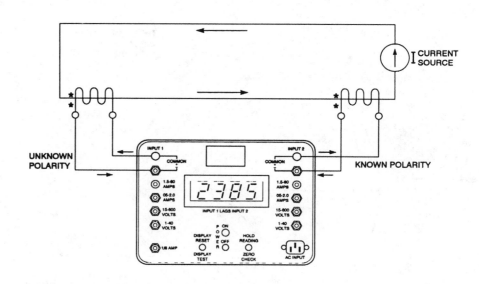

The phase-angle meter is designed specifically for power system applications, and may be used in the shop, field, or laboratory. It will measure and display the phase-angle relationship between two single-phase AC signals applied to its input circuits. The AC signals measured may be two voltages, two currents, or a voltage and a current. Current inputs are applied to a white common binding post and the appropriate black binding post. Potential inputs are applied to a white common binding post and the appropriate red binding post.

The digital display will indicate the phase angle, in degrees, that the signal applied to INPUT 1 lags the signal applied to INPUT 2.

REVIEW QUESTIONS

1. In AC circuits, why don't voltmeter and ammeter readings, multiplied together, equal watts?

2. What exactly does *theta* mean?

3. How does theta affect the efficiency of a circuit?

4. Explain what *class* means in a kilowatt-hour meter.

5. What does *Kh* mean in a kilowatt-hour meter?

6. Is a kilowatt-hour meter accurate enough for measuring watts?

7. Since typical phase-angle meters only have two inputs, how are three-phase circuits measured?

8. On a phase-angle meter connected so current readings lag the potential, what range of readings would I expect if the measured circuit had a leading power factor?

9. A kilowatt-hour meter for a three-phase, four-wire, Y circuit is miswired. A stopwatch check shows backward registration is the same as the corrected forward rotation at unity power factor. What was the original theta?

10. When a phase-angle meter shows current lagging potential by 30°, what does that mean in real world time?

Troubleshooting and Testing

In industry, few loads are entirely inductive or capacitive. Motors and transformers include wiring that has both a capacitive and resistive effect on the circuit. As a result, worst case inductive loads are seldom below 60 percent power factor (53°, lag).

Capacitive loads such as synchronous motors are rarely measured at less than 85 percent power factor (32° lead). When worst cases of power factor occur, it is usually because capacitor banks, normally used to improve power factor, are in the circuit when no other loads are on the line.

Typical voltage and current readings taken by a test technician on the three-phase, four-wire, Y transformer rated meter installation might give the following results.

Voltage readings from the three potential transformers feeding the meter read:

PHASE	VOLTAGE
A to N	122 Volts
B to N	121 Volts
C to N	120 Volts

Current readings from the three current transformers feeding the meter read:

PHASE	CURRENT
A	3.5 Amps
B	3.2 Amps
C	3.2 Amps

A phase-angle meter connected so current readings lag the voltage gives the following results:

PHASE	PHASE ANGLE
A	330 Degrees
B	332 Degrees
C	331 Degrees

The Kh of this meter is 1.8.

QUESTIONS
1. Is this load capacitive or inductive in nature?
2. What is the power factor of this circuit?
3. What watts would you expect to find if you took a stopwatch check on this meter installation?

ANSWERS
1. Capacitive. Any time the phase-angle meter is connected so readings show current lagging the voltage, readings of more than 270° indicate a capacitive load.
2. Add the three-phase angles and divide by three to come up with their average. (993/3 = 331.) Using a pocket calculator, read the cosine of 331°. If no calculator is available, look up the cosine of 360° minus 331, or 29°. Both answers are the same, 87 percent power factor (.87 × 100 = 87 percent).
3. Add all voltages, and divide by three to find average voltage (363/3= 121 volts). Add all currents, and divide by three to find average current (9.9/3 = 3.3 amps). Apply three-phase formula for watts for a Y circuit:

$$3 \times 121 \times 3.3 \times 0.87 = 1{,}042.2$$

A stopwatch check would take 62 seconds for 10 revolutions. Using the formula shown in Chapter 7 would indicate:

$$\frac{10 \times 1.8 \times 3{,}600}{62} = 1{,}042$$

Motor, Cable, and Transformer Test Equipment

8

INTRODUCTION

Phase rotation is important to ensure that three-phase motors turn in the proper direction. Phase sequence indicators provide an inexpensive means of verifying that phase rotations of energized supply circuits match those required for specific motor rotation.

Phase and motor rotation test sets are somewhat different from the phase sequence indicators mentioned above. The test set is designed to identify disconnected motor test leads. The test set is also capable of determining transformer polarity and circuit continuity.

Often, because of water damage, aging, or overheating, motor insulation tends to deteriorate. The megohmmeter has the ability to read out the resistance and evaluate the condition of the equipment.

Quite often, the ratio of a power transformer must be verified. One of the most efficient methods is by using a transformer's turns ratio test set.

OBJECTIVES

After studying this chapter, the student should be able to:

- *Understand why phase rotation of equipment is so important.*
- *Show what needs to be done to change phase rotation.*
- *Understand how to use the phase sequence indicator.*
- *Explain the operation of the phase sequence motor rotation indicator test set.*
- *Understand the operation and results obtained with the megohmmeter.*
- *Explain how to read and interpret the readings taken with the turns ratio test set.*

PHASE SEQUENCE INDICATOR

The phase sequence indicator is a rugged, lightweight instrument, used to determine the phase sequence or phase rotation of a three-phase system. It also detects open phases. The unit is actually a small three-phase motor, as shown in Figure 8–1. As such, the direction it rotates is determined by whether phase rotation is ABC or CBA.

Connection

When updating feeder cables, or replacing polyphase motors, the indicator ensures that proper phase rotation is maintained. Measuring rotation before and after installation ensures that equipment will not rotate backward.

Three color-coded leads with insulated clips are permanently attached to the instrument. The technician connects the red, white, and blue leads to A, B, and C phases. If conductors are not marked or color coded, select some system, such as left, center, and right. It is important to be consistent and use the same measuring sequence before and after connections of motors or equipment are made.

Should the device rotate clockwise when the protector switch is pressed, ABC phase rotation is indicated; if counterclockwise, the system is CBA.

Failure to rotate indicates an open phase.

Besides ensuring the proper rotation of motors, phase rotation meters are regularly used in power measurement installations. Reactive

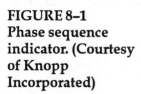

FIGURE 8–1
Phase sequence
indicator. (Courtesy
of Knopp
Incorporated)

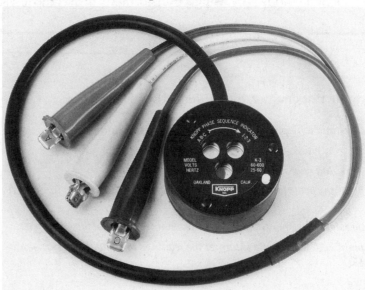

power, explained in Chapter 7, requires that monitoring meters be installed with a specific phase rotation, in order to properly register.

Lightweight units are also available that use a system of lights to indicate phase rotation and identify open phases.

MOTOR ROTATION INDICATOR

More sophisticated than a phase sequence indicator, the phase rotation indicator can also identify leads from de-energized motor coils.

Application

A motor rotation indicator is a test set used to identify the leads of a disconnected polyphase motor, so that when connected in the system, it will run in the desired direction. It is also used to identify phase rotation ABC (or with procedure modification, CBA) of energized AC power lines up to and including 600 volts. Other uses include determining transformer polarity and testing circuit continuity. The complete tester is shown in Figure 8–2.

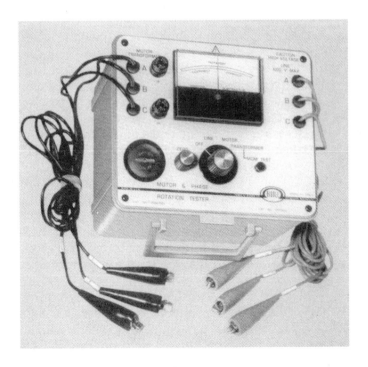

**FIGURE 8–2
Phase and motor
rotation tester.
(Courtesy of AVO
Biddle Instruments)**

The motor rotation indicator also provides, in a single instrument, facilities for identifying phase and polarity of winding sections of a multiple-winding motor.

CONTROLS AND CONNECTORS

This unit can be used on voltages up to 600 volts. Please use caution and proper protective equipment when connecting.

LINE LEADS—Three red (line) leads (marked A, B, and C), attached to the right of the panel, are for connection to an energized AC power system of up to and including 600 volts. Each lead is terminated in a battery clip and protected with a red rubber boot, which is identified as A, B, or C at the instrument panel, and again at the clip. The leads are permanently attached to the panel assembly, but may be readily replaced when necessary.

MOTOR LEADS—Three smaller black (motor) leads on the left— marked A, B, and C—are for connection to de-energized equipment. These battery clips are smaller and are identified in the same manner as the line leads.

SELECTOR SWITCH—A four-position switch is used to turn the set on and select the desired circuit.

- In the OFF position the meter and the battery are disconnected from all circuits.
- In the LINE position the meter is connected in the phase sequence circuit, and the LINE leads may be used to make a phase sequence measurement. The battery remains disconnected.
- In the MOTOR position the meter is disconnected from the phase sequence circuit and connected in the motor rotation circuit. The battery is also connected. The MOTOR leads may now be used for a rotation test.
- In the TRANS position, the meter and battery are connected so that the motor leads may be used for a transformer polarity test.

ZERO ADJUST—When the selector switch is in the MOTOR position, and the MOTOR leads are connected, this gives control of the meter deflection and may be used for balance in motor rotation tests.

TEST ON RELEASE—A push switch is connected in series with the battery. This opens the battery circuit when it is depressed.

METER—The meter is a zero-center instrument, with an ungraduated scale marked CORRECT on the right side and INCORRECT on the

---

WARNING
Never connect the black leads to an energized circuit.

left. Below the scale, the meter is marked SUBTRACTIVE on the right and ADDITIVE on the left.

Measurement

Three basic test circuits are incorporated in the tester:

- Uncalibrated bridge, used to determine direction of motor rotation.
- Phase rotation indicator.
- Polarity test circuit for transformer testing.

A single zero-center milliammeter serves as an indicator for all three circuits. A switch permits the selection of the desired circuit.

The source of energy for tests on de-energized equipment is a single flashlight battery. Fuses are in series with the motor A and C test leads as a protection for the operator, in case leads are accidentally touched to an energized circuit.

MOTOR ROTATION THREE-PHASE, THREE-TERMINAL

To check for proper motor rotation, set the selector switch to the motor position. Connect the black leads to the three terminals of the motor in any order. Operate the zero adjust control to set the meter pointer at the center of the scale. Manually turn the motor shaft slightly in the desired operating direction, while observing the meter. The meter will deflect ("kick") in one direction, and then in the opposite direction.

> **NOTE:** The first direction is significant. Ignore the second or opposite direction.

If the first deflection is in the incorrect direction, reverse any two motor leads. When the meter shows correct deflection, tag the motor leads according to the motor lead markings (A, B, C).

Should it be difficult to clearly distinguish the first deflection, because of a large second deflection, move the rotor a few degrees and repeat the test.

The initial deflection will always be in one direction, corresponding to a given armature rotation, which can be distinguished from possible false symmetrical oscillations. These oscillations can occur if residual magnetism is present in the armature teeth. The false oscillations can be observed, even without the presence of a dry cell, or if the "test on release" switch does not make contact.

PHASE SEQUENCE, THREE-PHASE

Set the selector switch to Line Position. Connect the red line leads to the three terminals of the line in any order. Observe the meter. If the meter reads incorrect, interchange any two of the line leads. When the correct reading is obtained, tag the line terminals according to the line lead markings (A, B, C). Phases A, B, C now follow in the true sequence.

Motor to Line Connection

When the motor terminals A, B, C are connected to corresponding line terminals A, B, C—which were tagged in true phase sequence as described above—the motor will run in the direction in which it was turned. To conform with NEMA MG1-2, 20, the numbers 1, 2, and 3 may be used in place of A, B, and C. Note that there is no standard rotation for polyphase induction motors. To facilitate arrangement of leads, two alternate methods of connection are tabulated below:

Motor		Line	Motor		Line
A	to	B	A	to	C
B	to	C	B	to	A
C	to	A	C	to	B

Transformer Polarity, Single-Phase

Set the selector switch to the Off position. Connect two adjacent high- and low-voltage transformer terminals, using a suitable jumper. Connect the C motor lead to one of these terminals. Connect the B motor lead to the remaining high-voltage terminals. Connect the A motor lead to the remaining low-voltage terminal. Set the selector switch to Trans position. Press the Test on Release button, and release. Observe the meter on release. Deflection of the meter indicates transformer polarity. Read either subtractive to the right, or additive to the left. If sensitivity is not adequate on low ratio transformers, switch to Motor position without changing leads, operate Zero Adj, control to set pointer at center, then test as above.

When two adjacent terminals of adjacent coils are connected together, the induced coil voltages or the terminal voltages will add in an additive transformer and subtract in a subtractive transformer. Figure 8–3 shows test set circuit for the transformer polarity test.

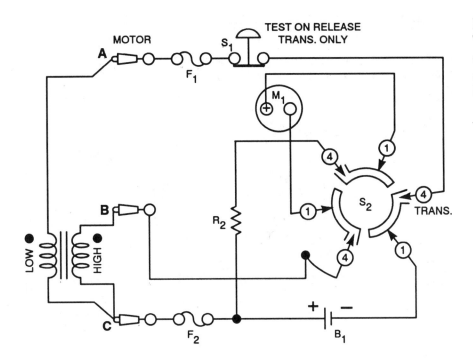

FIGURE 8–3
Transformer polarity
circuit. (Courtesy of
AVO Biddle
Instruments)

Continuity

Set the selector switch to Motor position. Connect A and B motor leads
together. Operate Zero Adj control until meter reads full scale. Use A
and B and motor leads for testing continuity. Resistances in the order of
ten ohms or lower will not appreciably reduce the full scale reading of
the meter. A resistance of 600 ohms will give approximately half-scale
deflection. Resistance above 20,000 ohms will show practically zero
deflection.

MEGOHMMETERS

Ordinary ohmmeters cannot be used for measuring resistance in multi-
millions of ohms, such as would occur in measuring that of conductor
insulation.

FIGURE 8–4
Megohmmeter
schematic diagram.
(Courtesy of AVO
Megger Instruments
Limited)

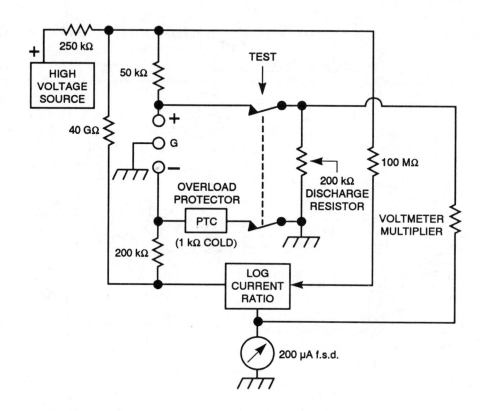

Application

To test adequately for insulation breakdown, it is necessary to use a much higher potential than could be furnished by the low-voltage, battery-powered multimeters. Instead, an instrument called a megohmmeter is used. Figure 8–4 shows the circuit of a typical model. Potential is measured between the conductor and the outside surface of the insulation. Any current flow calculated by Ohm's Law into resistance is read on the megohms scale of the meter.

In calculating the resistance to expect from testing insulation, one rule of thumb is: 1 megohm for each kV of voltage applied + 1 megohm.

Example: 1,000 volts applied should equal at least 2 megohms of resistance.

Voltage applied would depend on the voltage rating of the equipment measured. One of the most important aspects of testing insulation is keeping records. Any drastic change in resistance values will show up during testing.

The tester may be used for routine insulation resistance measurements on many components, such as motors, generators, transformers, high-voltage insulators, power cables, and wiring installations. Because it is battery operated, it is particularly suitable for service and field applications.

The tester can be employed in the routine maintenance of plant and installations to make measurements showing the state of the insulation when in service, and the amount of degradation that takes place due to the effects of corrosion, dirt, grease, moisture, etc. From such results, the future performance of these items can be estimated, and interruptions or breakdowns avoided.

Step-voltage tests, whether these are at the full rating or below, can be used to indicate the condition of the insulation subjected to moisture and dirt, or to a point of weakness. A ratio of 1:5 in the test voltage is suitable, and if the tests are carried out for a similar period of time, a difference between the resistance readings of the lower voltage test and higher voltage test points to a weakness in the insulation. Absolute values of insulation resistance are not of great importance, but the change in resistance is indicative of the trend in the insulation condition. Such tests are valuable when performed regularly on a scheduled basis and a record of results is kept.

Dielectric absorption testing, which involves insulation tests on capacitive circuits for several minutes, is also possible. The safety of electrical installations and apparatus depends upon the condition of the insulation, and it is essential that this is thoroughly checked when new equipment is installed. The test voltage applied should be high enough to break through any mechanical flaws arising from manufacture or installation. The tester can also be used to monitor the improvement in insulation within motor, generator, and transformer windings, etc., that result from drying-out procedures, following their service in excessively humid atmospheres.

When using this test equipment, certain procedures must be observed to ensure safety for the test technicians.

1. The equipment to be tested must be de-energized. Before making insulation tests, disconnect and isolate apparatus that is the subject of the test.
2. Before making any insulation measurements, the battery charger power supply lead MUST be removed from the instrument, and the battery charger socket cover replaced.
3. In connecting the instrument to the circuit, care must be taken to ensure that there is no hazard to the user, due to the voltage produced at the instrument terminals, which can be up to 5 kV.

4. The instrument generates very high voltages that, when applied to capacitive loads, such as long lengths of cable, will usually produce a lethal charge within half a second of pressing the TEST push-button. If the capacitive load becomes disconnected during a test, it will be left in a charged state. Care must be taken to avoid this. Also, when testing capacitive circuits, check that the meter reads less than 50 volts before disconnecting the instrument after carrying out a test.

5. For prolonged tests (e.g., on capacitive circuits), the TEST push-button may be locked down by pressing and then turning it clockwise with the thumb or forefinger. Care should be taken, with the instrument in this operating mode, that no harm or damage is done if it is left unattended, because it produces a dangerously high voltage. To prevent excessive discharge to the point of causing possible damage to the battery, it is advisable to check the instrument every 15 minutes.

For optimum safety, never lock the push-button unless it is absolutely necessary.

6. Care should be taken to ensure that excessive voltages are not applied to the terminals. The maximum continuous voltage that may be applied (from an external source) between the "+" and "–" terminals is 1,000 V, and between the "G" and "–" terminals is 250 V. If the meter indicates a voltage > 1,000V, the source of voltage must be removed as soon as possible (5 kV applied between "+" and "–" will cause damage after approximately 5 seconds). Maximum permissible current injected into the terminals from a high impedance source is 2 mA.

PRELIMINARY CHECKS

Using the diagram shown in Figure 8–5, perform the following preliminary checks.

1. Check that the meter pointer rests over the "∞" mark on the upper scale, with nothing connected to the terminals, and the TEST push-button is not pressed. If necessary, set the pointer with the mechanical adjuster situated below the meter, using a small screwdriver.

2. Turn the test voltage selector switch to the battery check position. Press the TEST push-button, and observe the meter reading. The pointer should come to rest very close to the "∞" mark on the scale. If it does not, leakage is occurring between the test leads, test clips, or possibly the instrument terminals. Release the push-button and check to find the leakage source.

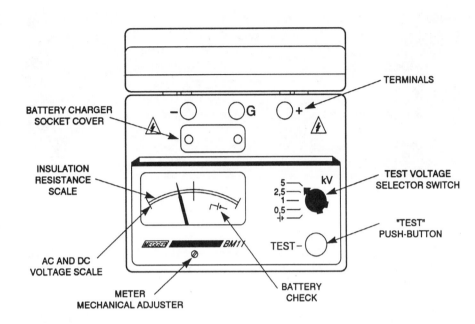

**FIGURE 8–5
Megohmmeter
controls. (Courtesy
of AVO Megger
Instruments Limited)**

3. With the test leads still connected to the instrument, join together the "+" and "–" test lead clips. Press the TEST push-button, and check that the meter reading is zero on the upper scale. Release the push-button. If the reading is not zero, the test leads are suspect and should be inspected for a fault (i.e., open circuit).

MAKING AN INSULATION TEST TO EARTH

1. Perform the preliminary checks as given above.
2. Connect the "+" test lead to the equipment to be tested, and the "–" test lead to Earth (frame of the equipment, etc.).

NOTE: a. If the equipment on test is not de-energized, the voltage present will be automatically indicated on the lower scale of the meter. Do not press the TEST push-button, but switch the circuit off, and ensure that no voltage is present before proceeding with a test.
b. If the equipment to be tested is known to be de-energized, yet a voltage reading is obtained, this is indicative of the presence of AC interference currents caused by capacitive or inductive coupling to live circuits.

FIGURE 8–6
Megohmmeter scale.
(Courtesy of AVO
Megger Instruments
Limited)

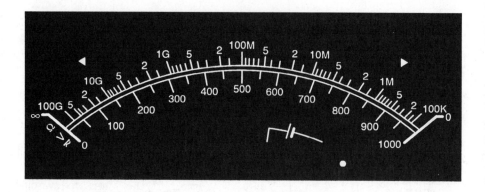

c. Although it is normal to connect the "–" terminal to Earth, the "+" terminal might equally well be used. Accuracy of measurement will not be affected.

3. With the correct test voltage selected, press the TEST push-button.
4. Read the value of the insulation resistance directly from the upper scale, as shown in Figure 8–6.

   NOTE: If, when taking a reading, the pointer displays small oscillations, this is an indication that the battery is becoming exhausted. When the pointer oscillations become large, the battery must be recharged before any more tests are carried out.

5. Release the push-button, and wait until the meter voltage reading is less than 50 V (shown on the lower scale) before disconnecting the test leads.

MAKING AN INSULATION TEST BETWEEN WIRES

1. Perform the preliminary checks as previously given.
2. Connect one test lead to each conductor, and then proceed as before in steps 3, 4, and 5 in Making an Insulation Test to Earth.

USING THE GUARD TERMINAL

For basic insulation tests, the guard terminal will not be used. Most insulation tests can be performed by connecting the specimen between the "+" and "–" terminals. These tests will show up any deficiencies in the insulation, whether they are caused by leakage through the insulator body or across its surface.

To distinguish between body leakage and surface leakage, the guard terminal "G" may be used. In this way surface leakage current is removed before it enters the measurement circuit via the "–" terminal.

In cable testing, there may be a path of leakage across the insulation between the bared cable and the external sheathing, perhaps due to the presence of moisture or dirt. Where it is required to remove the effect of this leakage, particularly at high testing voltages, a bare wire may be bound tightly around the insulation, and connected via the third test lead to the guard terminal "G," as shown in Figure 8–7.

Since the leakage resistance is effectively in parallel with the resistance to be measured, the use of the guard causes the current flowing through the surface leakage to be diverted from the measuring circuit. The tester, therefore, gives more nearly the true insulation resistance.

When the guard terminal is connected, it will cause two other errors, which, although usually negligible, can affect the measurement if the surface leakage is approximately <20 MΩ. These are:

- The leakage resistance has a loading effect on the voltage at the "+" terminal, which causes the actual voltage at the terminal to be slightly less than the voltage sensed by the reference circuit and results in a measurement error.
- The leakage current shunts a small proportion of current away from the measuring circuit and results in a measuring error.

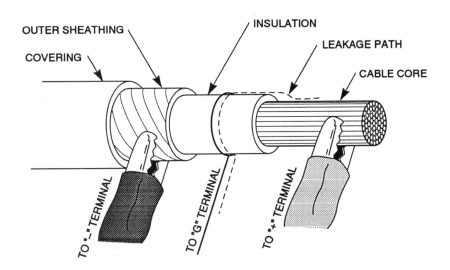

FIGURE 8–7
Cable testing.
(Courtesy of AVO
Megger Instruments
Limited)

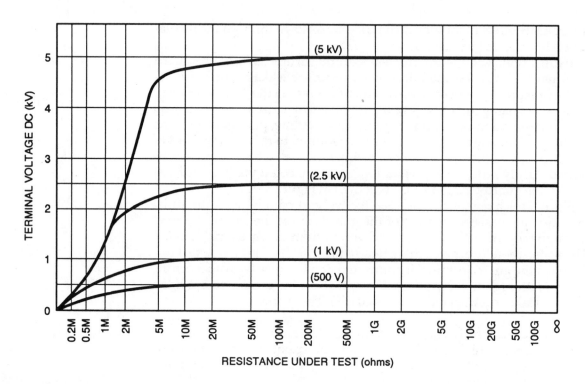

FIGURE 8–8
Typical terminal voltage characteristics. (Courtesy of AVO Megger Instruments Limited)

A 20 MΩ surface leakage resistance will produce a typical error of 0.5 percent of scale length. Serious errors will only occur if the surface leakage resistance is so low that the power available is insufficient to generate the required voltage across the surface leakage resistance. Refer to the graph of voltage characteristics, Figure 8–8.

PREVENTIVE MAINTENANCE

It is good practice to make regular tests of the insulation resistance of all larger machinery, and thus detect any incipient faults. When the tests are entered in the logbook, a considerable variation between test results will be noted. It is therefore important to test under similar conditions each time and to note the current weather status.

Damp weather—or damp conditions of use or storage—can cause large reductions in insulation resistance. Drying out by heat or by operation for a period should give a more consistent and appropriate insulation resistance value.

A counter effect to this occurs because the insulation resistance of the varnishes used in the construction of machine windings becomes lower when hot than when cold. Therefore, for constant comparisons, the temperature of the machine under test should also be noted.

The best plan is to make regular insulation checks as soon as possible after the machine has been shut down. The insulation resistance is then likely to be at its lowest operational value; this becomes the figure that would show any continuing mechanical depreciation or potential insulation breakdown.

If the machine stands idle in humid conditions, a worse picture might well apply, but this would normally be assumed to be safe during the running up to temperature, provided that the resistance at working temperature remains unchanged.

TESTING MOTORS AND GENERATORS

1. Disconnect the equipment from the line by opening the main switch and removing the main fuses.
2. Join together both terminals on the motor side of the double pole main switch.
3. With a contactor operated starter, where all the lines to the motor are disconnected at off position, it is necessary to make tests to Earth on both the incoming and outgoing terminals of the starter.
4. Connect the "−" terminal of the BM11 MEGGER tester to Earth, using the frame of the motor or switch.
5. Using the "+" terminal, measure the resistance of each part of the circuit in the usual way. If the value is unsatisfactory, then separate tests in starter, motor, and cables must be carried out to locate the defect.
6. If the motor itself is suspect, disconnect the supply cables and, with one lead connected to the frame, carry out the following tests.
7. Test with the armature and field windings connected together.
8. Test with the brushes lifted from contact with the commutator.
9. Test on the armature only, section by section.
10. If all resistances are low, the fault can usually be remedied by complete and careful cleaning of the machine.

Equipment that has been in service for a while can accumulate metallic or other conducting dust, especially when mixed with oil from bearings, etc. The leakage paths from such deposits are eliminated by thorough cleaning.

CIRCUIT DESCRIPTION

As shown in Figure 8–4, there are two main sections within the tester: the inverter, producing the high test voltage, and the current measuring circuit. The high voltage "+" terminal is positive with respect to the guard "G" and "–" (measurement) terminals.

The inverter operates in three modes:

- Constant output voltage for low loads; i.e., resistances >10MΩ.
- Constant output power for heavy loads; i.e., resistances approximately 5 MΩ.
- Constant output current for very heavy loads and short circuit.

The measurement circuit uses a pair of matched transistors to produce a voltage proportional to the logarithm of the ratio of leakage current (current through test sample) to applied voltage.

The scale shape is of the basic form $\frac{1}{5} \log_{10} \frac{100}{R}$, where R is the resistance of the test samples in MΩ. Deliberate cramping at each end of the scale enables the 0 and ∞ marks to be shown and still give five decades of resistance measurement.

The maximum allowable voltage that may be applied to the test terminals is shown in Figure 8–9. The maximum current through all terminals is 2 mA.

TRANSFORMER TURNS RATIO TEST SET APPLICATION

Turns ratio testing provides a thorough measure of all portions of transformer performance. It is a direct means to detect loose or open connections, shorted or open turns in the transformer windings, incorrect polarity connections, and to verify precisely the no-load voltage ratio of the transformer at all tap positions.

**FIGURE 8–9
Maximum allowable
voltage. (Courtesy of
AVO Megger
Instruments
Limited)**

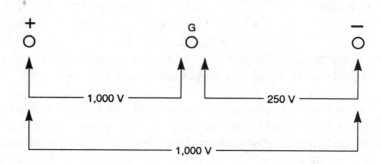

The turns ratio of a transformer is defined as the number of turns in one winding, in relation to the number of turns in the other winding of the same phase, and is equal to the no-load voltage ratio. Defined mathematically:

$$\frac{E_p}{E_s} = \frac{N_p}{N_s} = TR$$

E_p = Voltage primary
E_s = Voltage secondary
N_p = Number of turns primary
N_s = Number of turns secondary
TR = Turns ratio (primary to secondary)

For reliable operation, and parallel connections of power and distribution transformers, it is extremely important that the turns ratio be precisely determined. The turns ratio is required to be within ±1/2 percent (.005) of the indicated nameplate voltage ratio.

Nameplate voltage ratio (N.P.) of a transformer is defined as the line-line voltage of the primary winding to the line-line voltage of the secondary winding at no load.

$$\frac{\text{H.V. (primary) line–line voltage}}{\text{L.V. (secondary) line–line voltage}} = \text{N.P. or } \frac{E_p}{E_s} = \text{N.P.}$$

Since the no-load voltage ratio is directly proportional to the turns ratio, an accurate measurement of the turns ratio will provide the no-load voltage ratio without having to test with dangerously high voltages.

When primary voltage is compared to secondary voltage on a potential transformer, the no-load voltage ratio is nearly equal to the turns ratio. Any difference between the two ratios is caused by a voltage drop in the winding that results from the magnetizing current flowing through the winding. Normally this difference is less than 0.1 percent, or .001. By exciting a transformer winding with a known voltage and applying this same voltage to a reference transformer (in which the exact turns ratio is known), a balance circuit can be used. When both the transformer under test and the reference transformer voltage ratios are equal, the turns ratio is obtained.

Transformer turns ratio test sets are designed to operate on the above principle. In addition, the test set is designed so that, during the ratio tests, polarity, open or short-circuited turns, and the vector relationships of the various transformer windings are checked. Several differ-

FIGURE 8–10
Transformer turns
ratio test set.
(Courtesy of AVO
Biddle Instruments)

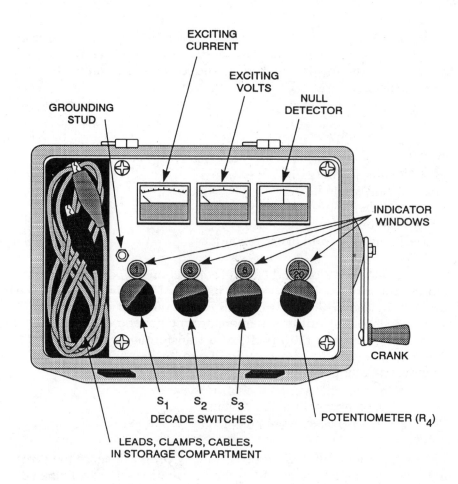

ent types of turns ratio test sets are manufactured; however, they generally operate by the same principle. A typical turns ratio test set is shown in Figure 8–10.

Measurement

Figure 8–11 shows a simplified schematic diagram of a test set, which will be used to explain the theory of operation. The test set is arranged so that the transformer to be tested and the adjustable ratio reference transformer in the test set are excited from the same source voltage. The secondary windings are connected in a series opposing configuration through a null detector. When the ratio of the reference transformer is adjusted so no current flows in the secondary circuit (null point), the voltage ratios of the two transformers are equal. Since no load exists on either secondary, and the no-load voltage ratio of the reference trans-

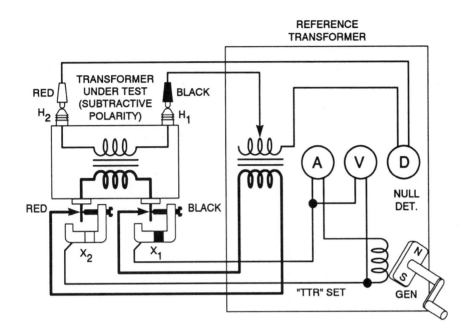

FIGURE 8-11
Simplified
schematic diagram.
(Courtesy of AVO
Biddle Instruments)

former is known, the voltage ratio of the transformer under test is obtained; thus its turns ratio is also known.

In order to have a reference transformer that is adjustable, and have a continuous known voltage ratio, the test set's reference transformer has a set reference winding as in Figure 8–12. Since the number of turns in the reference transformer winding is 90 turns, and the voltage applied to this winding is 8 VAC, the volts per turn is equal to:

$$\frac{8 \text{ VAC}}{90 \text{ Turns}} = .088 \text{ V/T}$$

This known relationship means nothing in itself, until the adjustable winding is also excited by some level of voltage. Since this voltage level is unknown, and depends entirely upon the voltage ratio of the transformer under test, the reference transformer is designed so an adjustment of its turns ratio is possible.

Figure 8–13 shows an example of how the test set operates. With the reference transformer winding excited by the 8 VAC, and the low-voltage winding of the transformer under test excited by this same 8 VAC, a voltage is developed on the high voltage of the transformer under test, which is directly proportional to its voltage ratio. In Figure 8–13, the transformer under test is assumed to have a 2:1 ratio; thus the voltage developed on its high-voltage winding is 16 VAC.

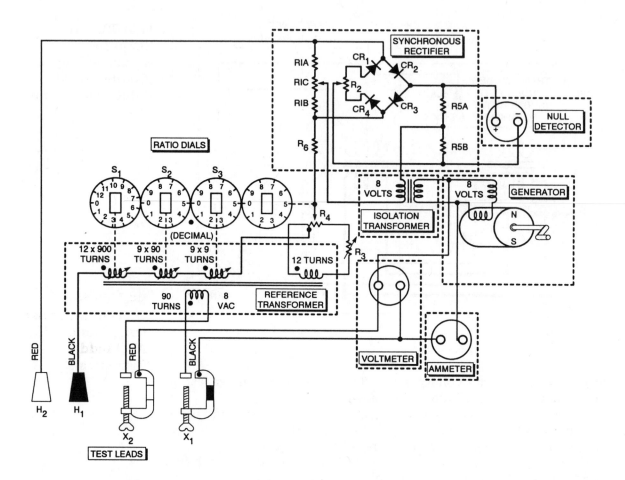

FIGURE 8–12
Schematic diagram.
(Courtesy of AVO
Biddle Instruments)

$$2 \times 8 \text{ VAC} = 16 \text{ VAC}$$

This 16 VAC is applied to the reference transformer adjustable winding. With all decade switches (S_1, S_2, S_3, R_4) at zero, the volts per turn is equal to:

$$\frac{16V}{1 \text{ Turn}} = 16 \text{ V/T}$$

In order to null the test set, the reference transformer must be adjusted so the turns ratio of the reference transformer is equal to the turns ratio of the transformer under test.

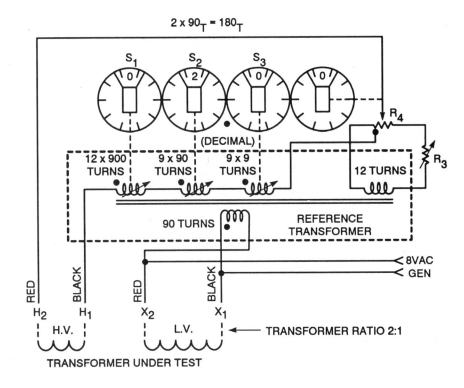

FIGURE 8–13 Transformer under test. (Courtesy of AVO Biddle Instruments)

Since the voltage ratio is:

$$\frac{16\,(E_p)}{8\,E_s} = 2\frac{16}{8}$$

The reference transformer turns ratio must be adjusted to:

$$\frac{x}{90T} = 2$$
$$x = (2)\,(90T)$$
$$= 180 \text{ Turns}$$

This shows the reference transformer must be adjusted by adding 180 turns.

By examining the decade switches, the final setting can be determined. Notice one position of the S_1 decade switch will add 900 turns, which exceeds what is required for balance. One position of the S_2 decade switch is equal to 90 turns; thus, at position 2, 180 turns have been added, which should null the test set null detector. With the null detector at balance, all decade switches except S_2 would be at zero, thus:

Reading		Turns Added
S_1 @ 0 =		0 Turns
S_2 @ 2 =		180 Turns
S_3 @ 0 =		0 Turns
R_4 @ 0 =		0 Turns
2 =		180 Turns

The final reading would then be 2, or 2:1 ratio. If a null condition is not obtained, the transformer under test would have some ratio other than exactly 2:1 ratio. With a tolerance of 1/2 percent, the actual ratio could be:

2 × .005	=	0.01
2 + 0.01	=	2.01 Upper limit
2 − 0.01	=	1.99 Lower limit

(R_4 is provided for fine adjustment and has a 1.2 turn per division scale.)

Figure 8–14 shows a 4:1 ratio transformer under test in which the same conditions still apply; however, 360 turns are now required to null the TTR test set.

This test set, along with most others, has limits in which it must operate, which are determined by the number of adjustable turns available in the reference transformer and the magnetizing current required by the transformer under test. The test set used in the examples so far has a limit of 130:1 ratio. This can be seen by multiplying the 90 turns of the reference transformer by 130, which is equal to 11,700 turns. Using all of the decade switches at their maximum settings provides the following:

Reading		Turns Added
S_1 @ 12	=	10,800 Turns
S_2 @ 9	=	810 Turns
S_3 @ 9	=	81 Turns
R_4 @ Max	=	10 Turns
29.999		11,701 Turns

Ratios as high as 330 can be measured by using auxiliary equipment. In addition, if the excitation current of the transformer under test is excessive (more than .8 amps @ 8 V), or the secondary voltage rating of

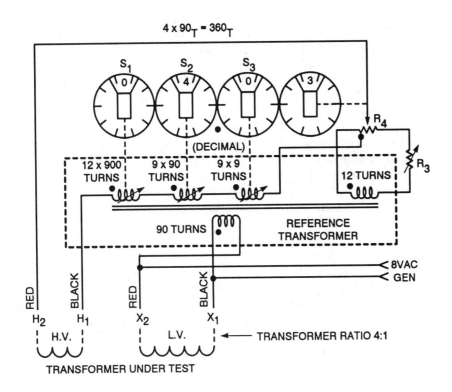

FIGURE 8–14
Transformer ratio
test. (Courtesy of
AVO Biddle
Instruments)

the transformer is less than 8 VAC, this same test set with auxiliary equipment can still be used. By exciting the primary of the transformer under test, the required excitation current required can be reduced; however, the turns ratio measured is the inverse ratio:

$$\frac{1}{2:1} = .5$$

$$\frac{1}{4:1} = .25$$

$$\frac{1}{130:1} = .00769$$

Connection

The transformer turn ratio test set must not be connected to an energized circuit. Correct connections for a test are extremely important, or improper operation and readings will result.

The set has four leads marked X_1, X_2, H_1, and H_2.

- The exciting lead (X_1) is used to connect the transformer under test low-voltage winding to the primary of the reference transformer in the TTR test set.
- The red exciting lead (X_2), non-polarity, connects the low-voltage winding to the non-polarity side of the primary of the reference transformer.
- Secondary lead (H_1), black, is used to connect the polarity side of the transformer under test high-voltage winding to the secondary polarity side of the reference transformer.
- Secondary lead (H_2), red, is used to connect the non-polarity side of the transformer under test high-voltage winding to the secondary non-polarity side of the reference transformer. Figure 8–12 shows this relationship.

The first step in determining the proper connections is to examine the transformer's nameplate polarity and/or schematic diagrams. The proper winding and polarity must be observed.

The 1∅ distribution transformer illustrated in Figure 8–15 shows that H_1 and X_1 are the instantaneous polarities of this transformer; thus the ratio of:

$$\frac{H_1 - H_2}{X_1 - X_2}$$

will be determined.

The H_1 and X_1 polarities must be observed in the following procedures:

- Black (X_1) of the test set is connected to the X_1 terminal of the transformer to be tested.
- Red (X_2) of the test set is connected to the X_2 terminal of the transformer to be tested. (This connection determines the instantaneous current flow in the transformer low-voltage winding from X_1 to X_2.)
- Black (H_1) of the test set is connected to the H_1 terminal of the transformer to be tested.
- Red (H_2) of the test set is connected to the H_2 terminal of the transformer to be tested. (This connection determines the instantaneous current flow in the reference transformer secondary winding.)

Notice that the instantaneous current flow in the high-voltage winding and low-voltage winding of the transformer under test are in opposite directions (subtractive). Figure 8–16 shows what occurs if

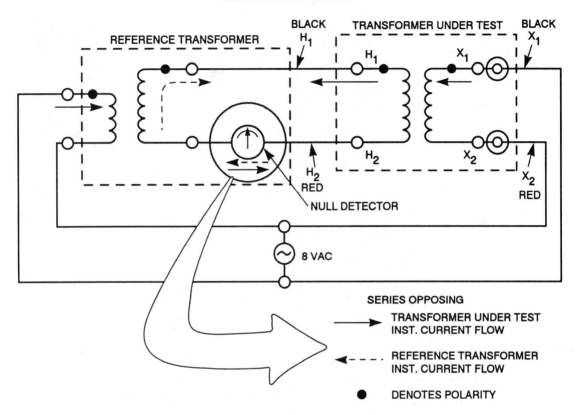

"SUBTRACTIVE POLARITY"

REFERENCE TRANSFORMER

BLACK
H₁

TRANSFORMER UNDER TEST

BLACK
X₁

H₁

X₁

H₂

X₂

H₂
RED

X₂
RED

NULL DETECTOR

8 VAC

SERIES OPPOSING

TRANSFORMER UNDER TEST
INST. CURRENT FLOW

REFERENCE TRANSFORMER
INST. CURRENT FLOW

● DENOTES POLARITY

FIGURE 8–15
Correct connection
for turns ratio test.
(Courtesy of AVO
Biddle Instruments)

improperly connected. The instantaneous current in the transformer under test is still in the opposite direction; however, the current flow in the reference transformer is now aiding, versus opposing. Remember, the test set reference transformer is connected for series opposing through the null detector. If the current flows in the same direction (aiding) through the reference transformer, the null meter will never be at its balanced point. Figures 8–15 and 8–16 show subtractive polarity transformer connections; however, not all transformers are subtractive. When testing an additive transformer, it is necessary to interchange the secondary leads H_1 and H_2 to properly connect the test set, Figure 8–17 (page 209).

With all four of the ratio dials set at zero, and the test set properly connected, a quarter turn of the generator hand crank indicates, by deflection of the null detector, both continuity and polarity. If these are

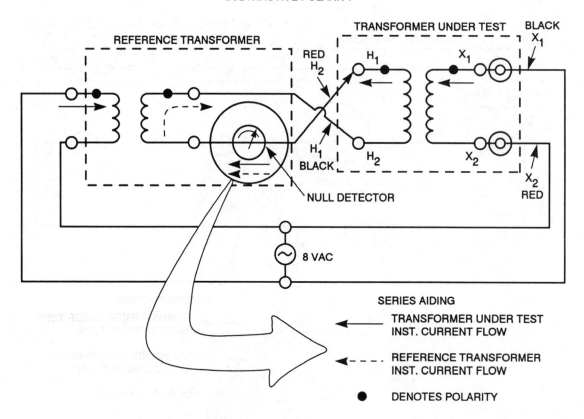

"SUBTRACTIVE POLARITY"

REFERENCE TRANSFORMER

TRANSFORMER UNDER TEST   BLACK

RED
H_2

H_1 X_1

X_1

H_1
BLACK

H_2 X_2

NULL DETECTOR

X_2
RED

8 VAC

SERIES AIDING

TRANSFORMER UNDER TEST
INST. CURRENT FLOW

REFERENCE TRANSFORMER
INST. CURRENT FLOW

● DENOTES POLARITY

**FIGURE 8–16
Incorrect connection
for turns ratio test.
(Courtesy of AVO
Biddle Instruments)**

correct, the pointer of the null detector will swing to the left. The voltmeter and null detector are mounted side-by-side, so they may be observed simultaneously.

Beginning with the dial switch on the left and turning the crank slowly at first, the ratio of the reference transformer is increased to equal that of the transformer under test. At or near balance, increase the hand generator speed, so the voltmeter point stands over the eight-volt reference mark on the scale. Fine tune to obtain balance, and record decade switch values.

SUMMARY

Small, hand-held, phase rotation meters are of two basic types. One is a miniature three-phase motor that responds by rotating clockwise or

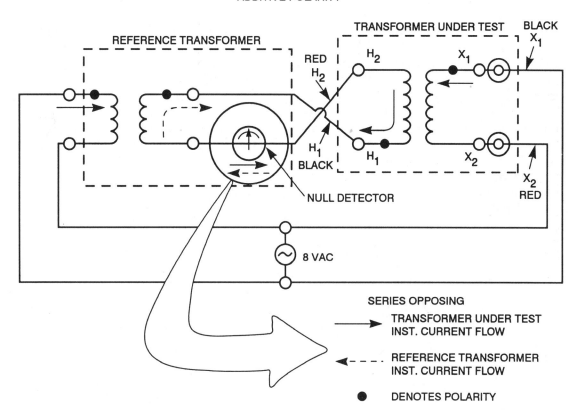

"ADDITIVE POLARITY"

FIGURE 8–17
Correct connection
for turns ratio test of
an additive
transformer.
(Courtesy of AVO
Biddle Instruments)

counterclockwise in the direction of the measured phase rotation. The second uses a series of pilot lights to indicate phase rotation and to identify live phases.

Phase and motor rotation test sets have the capability of measuring de-energized motor leads, to determine the wiring sequence needed to achieve a desired phase rotation. In addition, the test set can identify transformer polarity and verify circuit continuity.

As motors age, insulation values change. That is why periodic tests with a megohmmeter are important. By keeping ongoing records of test results, the state of insulation health can be tracked and monitored.

It is important to know that transformers have the correct ratio and polarity markings. The transformer turns-ratio test set incorporates all necessary circuitry in one unit to carry out these tests.

REVIEW QUESTIONS

1. Will the phase sequence indicator work on de-energized circuits?
2. How many leads must be changed in a three-phase circuit to reverse phase rotation?
3. If the conductors measured were unmarked, how could you decide which was A, B, or C phase?
4. How does the phase and motor rotation test set work on de-energized circuits?
5. Which leads are connected to energized circuits in the motor rotation test set?
6. What is the rule of thumb for calculating megohms in insulation testing?
7. Does the megohmmeter use AC or DC voltage?
8. What purpose does the ring guard serve when conducting insulation tests?
9. How would energizing the unknown transformer and test set from different sources affect the test?
10. Besides ratio, what are some of the other conditions checked by the transformer turns ratio test set?

Troubleshooting and Testing

A pad mount transformer that will feed a new addition in a factory complex is being installed. Prior to energizing, a transformer turns ratio test set is used to verify the unit. Ratio between the primary and secondary is tested not only to verify the correct turns ratio, but to ensure that it meets ANSI (American National Standards Institute) specification C57.12, which specifies that the ratio must be within 0.5 percent.

QUESTIONS
1. Using the injection methods described in this chapter, what would be the nominal readings of the ratio dials after the test is completed?
2. What are the minimum and maximum ratio readings that would still allow this transformer to meet ANSI C57.12 specifications of 0.5 percent?

ANSWERS
1. $\dfrac{12470}{277}$ = Ratio of 45.02
2. $45.02 \times 0.005 = 0.23$. Minimum ratio = 44.79
   Maximum ratio = 45.25

Circuit Breaker Test Sets 9

INTRODUCTION

In the last chapter, motor, cable, and transformer test sets were covered.

In this chapter, high-current, low-voltage tests will be covered, including their operation and use in circuit breaker overload testing. Available in many sizes and current ratings, these sets are useful whenever a variable high-current source is needed for testing. Some additional uses are current transformer ratio testing and panel ammeter calibration.

> **NOTE:** All the test sets described in this chapter are designed to test circuit breakers in de-energized circuits.

OBJECTIVES

After studying this chapter, the student should be able to:

- *Explain how the molded-case circuit breaker operates to protect against overloads.*
- *Understand long time characteristics of a breaker.*
- *Explain the instantaneous feature of a breaker.*
- *Know how to read manufacturer's curves for breaker performance.*
- *Identify the application of high-current circuit breaker test sets.*
- *Explain the different tests and measurements made with circuit breaker test sets.*

CIRCUIT BREAKER TEST SETS

Figure 9–1 shows a self-contained test set that consists of a variable high-current output and appropriate control circuits and instruments for testing thermal or magnetic, molded-case circuit breakers or motor overload relays. Figure 9–2 shows a simplified schematic of the unit.

The test set is capable of testing the time delay characteristics of devices rated up to 125 amperes, using an industry recommended test current of three times their rating (375 amperes). Additionally, to perform an instantaneous trip test, it will provide 750 amperes through a typical 125-ampere molded-case circuit breaker, when connected with the test leads provided with the test set.

**FIGURE 9–1
Circuit breaker test set. (Courtesy of AVO Multi-Amp Corporation)**

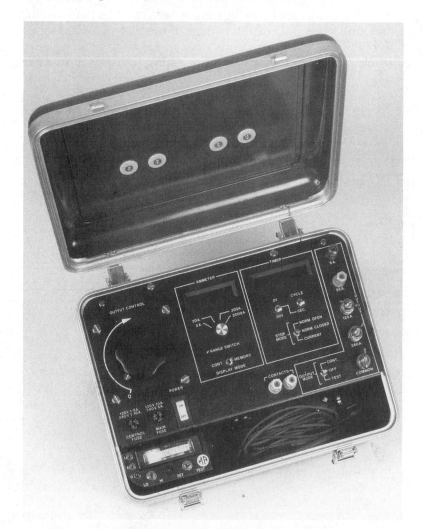

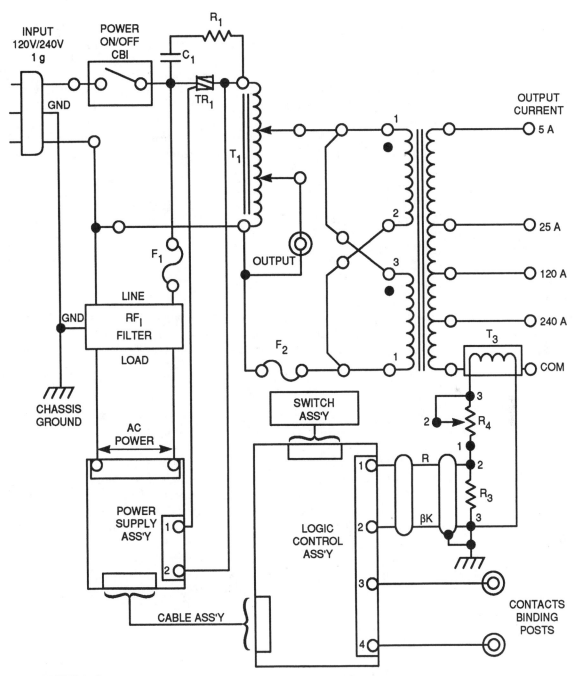

FIGURE 9–2
Circuit breaker test set schematic

The output circuit is designed to permit short time overloads. Output ranges will provide several times their current rating, provided the output voltage is sufficient to "push" the desired current through the impedance of the test device.

Impedance is the electrical characteristic in an AC device that is roughly equivalent to resistance in a DC device, in that both oppose the flow of current.

The output is continuously adjustable in four ranges, in order to accommodate a variety of test circuit impedances:

0–120 Volts at 5 amperes
0–24 Volts at 25 amperes
0–6 Volts at 120 amperes
0–3 Volts at 240 amperes.

Connections and Measurements

Molded-case circuit breakers and motor-overload relays should be tested to verify the time delay characteristics of the long and short time delay elements. They should also be tested for the minimum operating point of the instantaneous element, as it applies to the particular device being tested. The following paragraphs describe the measurement devices used to aid in the performance of testing.

Four output terminals, at various voltage and current ratings, are provided to adapt the test set to a wide variety of test circuit impedances.

All low-voltage, high-current test sets operate best by using the terminal with the highest-current, lowest-voltage rating suitable for the test. In this way, finer adjustments can be obtained by making full use of the variable autotransformer range. Even the smallest currents can be obtained from high-current terminals. Low-current, high-voltage terminals should be used only when testing high impedance devices where low-voltage terminals will not "push" the desired test current through the device. The operator should start with the lowest voltage terminal, and move to a higher voltage terminal only when necessary.

Output Initiate Circuit

The test set utilizes a completely solid-state output initiating circuit. Silicon controlled rectifiers (SCRs) are used instead of contactors to initiate the output, in order to increase reliability and eliminate contact maintenance.

Initiating circuits have both momentary and maintained modes to control output duration. The momentary mode is used whenever the output is to be on for a short time period. This might be when performing instantaneous trip tests, or to avoid damage or overheating of the device under test, while setting the test current.

In the maintained mode, the output remains energized until manually turned off. When performing timing tests, it remains on until the device under test operates—which both stops the timer and de-energizes the output.

To measure the output current, the test set uses a solid-state digital ammeter with multiple ranges. These are equipped with a "read-and-hold" memory, to measure short duration currents.

A solid-state digital timer measures the total elapsed time of the test in either seconds or cycles. A built-in crystal-controlled oscillator is used so that accuracy does not depend on line frequency.

A control circuit automatically starts the timer when the output is energized, and automatically stops the timer and de-energizes the output when the device under test operates. Switch selection allows the unit to operate in the following modes:

- Current actuated—Used to test a device that has no auxiliary contacts to monitor, such as a single-pole circuit breaker. The timer stops when the output current is interrupted.
- Normally closed—Used to test a device with normally closed contacts. The timer stops, and the output is de-energized when the contacts open.
- Normally open—Used to test a device with normally open contacts. The timer stops, and the output is de-energized when the contacts close.

Molded-case circuit breakers consist of switching and protective elements together in one unit, surrounded by an insulating material. Single-phase units rated at 15 amperes and 120 or 240 volts are found in many homes. Three-phase models, rated for high current and 600-volt circuits, are also available, as shown in Figure 9–3.

Molded-case circuit breakers consist of two separate elements. One is a set of contacts and mechanical linkage for manual operation of the breaker as a switch in an electrical circuit. The other element is a device that senses and reacts to an overload or short circuit. Normally, the time delay overload device is thermal, and the instantaneous overload device, when supplied, is magnetic. The thermal element ordinarily uses a bimetallic strip—two pieces of dissimilar material bonded together, as shown in Figure 9–4. An overload causes an increase in heat, which will result in moving the bimetallic unit and will eventually trip the

FIGURE 9–3
Three-phase
molded-case circuit
breaker

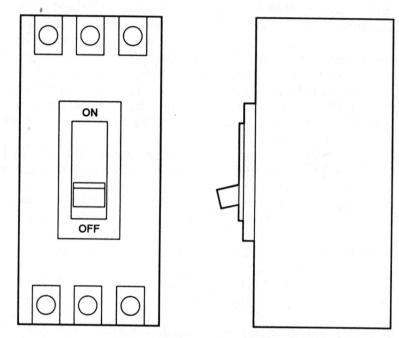

3 Ø MOLDED CASE CIRCUIT BREAKER

FIGURE 9–4
Thermal overload

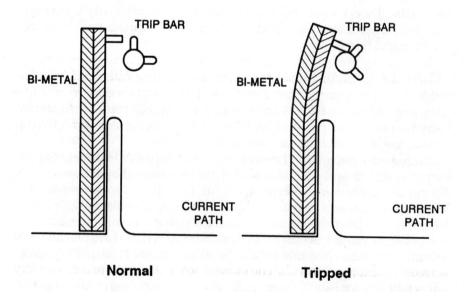

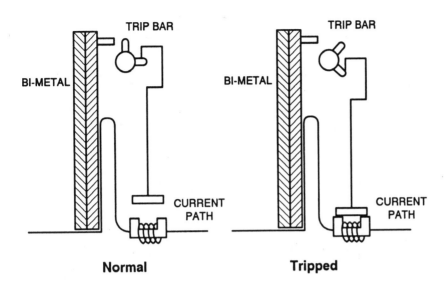

FIGURE 9–5
Instantaneous trip
magnetic element

circuit breaker. The magnetic element operates with no intentional time delay, to provide instantaneous protection against high magnitude faults, as shown in Figure 9–5.

READING A TIME CURVE

Figure 9–6 shows the manufacturer's time curves for a typical molded-case circuit breaker. The recommended test for overload is three times the rating of the circuit breaker.

Assume a molded-case circuit breaker rated at 125 amperes is being tested, and no magnetic adjustment is available. At the bottom of the chart, find the multiple of current rating column that is flush with the left edge of the chart. Look for a multiple of three (position A), which will be between the two and four. With a straight edge on three, look at the top, where the cross-hatched curves veer to the left. The location where the three multiple crosses this curve (position B) shows that this breaker can trip out in 40 seconds, but no later than 115 seconds at three times 125 amperes, or 375 amperes. Most manufacturers also warn that actual readings may vary from 10 percent to 25 percent of those numbers.

Instantaneous trip is taken at some multiple that is higher than the thermal trip. This time look for a multiple of current rating of four times breaker rating (position C). The curve shows that the breaker will trip anywhere from 0.01 to 0.04 seconds.

FIGURE 9–6
Typical time curve, molded-case circuit breaker. (The information contained herein is copyrighted information of the IEEE, extracted from IEEE Std 242–1975, copyright © 1975 by the Institute of Electrical and Electronics Engineers, Inc. This information was written within the context of IEEE Std 242–1975 and the IEEE takes no responsibility for or liability for any damages resulting from the reader's misinterpretation of said information resulting from the placement and context in this publication. Information is reproduced with the permission of the IEEE.)

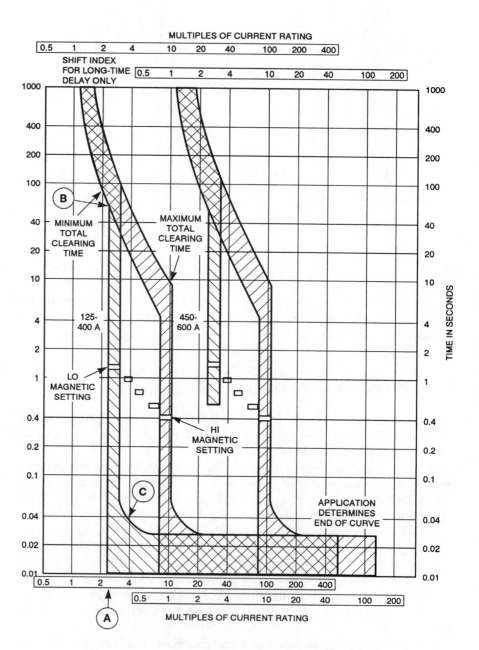

Measurement Procedure for Thermal Element of Molded-Case Circuit Breakers

Use the following sequence to test the breaker:

1. Set controls as follows:

    • POWER ON/OFF switch is in OFF position (instrument displays off).
    • OUTPUT CONTROL knob at minimum "0" position.
    • OUTPUT MODE switch in center off position.

2. Connect one end of a high-current lead to one pole of the circuit breaker. Connect one end of this lead to the common terminal of the test set.
3. Connect one end of the second high-current lead to other side of the same pole of the circuit breaker. Connect other end of this lead to output terminal labeled 240A.
4. Connect test set to suitable single-phase power supply.
5. Turn test set on with POWER ON/OFF switch (instrument displays should light).
6. Use RANGE switch to select ammeter range, so test current will be near full scale, and no less than 10 percent of full scale.
7. Put ammeter display mode switch in MEMORY position.
8. Place time STOP MODE switch in CURRENT position.
9. Select desired timer display mode and range.
10. Rotate OUTPUT CONTROL knob clockwise, momentarily press OUTPUT MODE switch on MOM., and release. Observe current reading retained by ammeter.
11. Continue to rotate OUTPUT CONTROL knob clockwise while jogging (repeatedly moving to MOM. position and releasing) OUTPUT MODE switch until desired test current is reached. Suggested test current is three times (3x) the rating of the circuit breaker.

    If desired test current is not reached with OUTPUT CONTROL knob at maximum clockwise rotation, return knob to zero, and transfer output lead from terminal labeled 240A to terminal labeled 120A. Proceed with current adjustment as in Steps 10 and 11.

    NOTE: Before starting test, allow time for the thermal element to cool; otherwise incorrect tripping time may result.

12. Put ammeter DISPLAY MODE in CONT. position.
13. Start test by moving OUTPUT MODE switch to MAINT. position.

> **NOTE:** Test current may decrease (fall off) during the test because the resistance or impedance of the test circuit increases as it heats up. Rotate OUTPUT CONTROL knob clockwise to keep test current at desired value.

14. When circuit breaker trips, timer stops and output is de-energized. Timer indicates total elapsed time of the test in seconds or cycles.
15. Turn test set off with POWER ON/OFF switch.
16. Record test results on Test Record Card.

> **NOTE:** Some types of circuit breakers are intended to trip only under high-current fault conditions, usually ten times (10x) the rated current. They have only instantaneous characteristics and therefore will not trip using the procedure described above.

Measurement Procedure for Instantaneous Element of Molded-Case Circuit Breakers

Use the following sequence to test the instantaneous element of molded-case breakers.

1. Control settings are as follows:

    - POWER ON/OFF switch is in OFF position (instrument displays off).
    - OUTPUT CONTROL knob at minimum "0" position.
    - OUTPUT MODE switch in center OFF position.

2. Connect one end of a high-current lead to one pole of circuit breaker. Connect one end of this lead to the common terminal of test set.
3. Connect one end of second high-current lead to other side of same pole of circuit breaker. Connect other end of this lead to output terminal labeled 240A.
4. Connect test set to suitable single-phase power supply.
5. Turn test set on with POWER ON/OFF switch (instrument displays should light).
6. Use RANGE switch to select ammeter range so test current will be near full scale, and no less than 10 percent of full scale.
7. Put ammeter DISPLAY MODE switch in MEMORY position.
8. Place time STOP MODE switch in CURRENT position.
9. Select desired timer display mode and range.

10. Rotate OUTPUT CONTROL knob clockwise, momentarily press OUTPUT MODE switch on MOM., and release. Observe current reading retained by ammeter.

    If desired test current is not reached with OUTPUT CONTROL knob at maximum clockwise rotation, return knob to zero and transfer output lead from terminal labeled 240A to terminal labeled 120A. Proceed with current adjustment as above.

11. Continue Step 10 until circuit breaker trips. Observe current reading retained by ammeter. Time indicates elapsed time of test in seconds or cycles.

12. Repeat test, starting with OUTPUT CONTROL knob at position just below trip current of instantaneous element observed in Step 11.

   NOTE: Before starting test, allow time for the thermal element to cool; otherwise incorrect tripping time may result.

13. When circuit breaker trips, timer stops and output is de-energized. Current reading is retained on ammeter. Timer indicates elapsed time in seconds or cycles.

14. Turn test set OFF with POWER ON/OFF switch.

15. Record test results on Test Record Card.

   NOTE: Refer to manufacturer's instructions for instantaneous trip time. If increasing test current does not decrease tripping time, current at which minimum tripping time was first obtained is the instantaneous trip current value.

MOTOR OVERLOAD RELAYS

The prime function of motor overload relays is to prevent operation of a motor for an excessive period of time when an overload condition exists. Motor overload relays incorporate an element that actuates a set of contacts connected to the motor control circuit. These contacts open the circuit of the holding coil in the motor starter and interrupt the power to the motor.

In general, motor overload relays operate by one of three principles:

- Thermal—By using heat to melt solder, which allows a spring to trip the relay.
- Thermally—Bimetallic strips (made out of two different metals) bend from the heat and trip the relay.
- Electromagnetic—Uses a solenoid device to pull in when currents are greater than normal. Figure 9–7 shows some typical examples.

FIGURE 9–7
Magnetic overload
relays. (Courtesy of
Allen Bradley
Company and
Square D Company)

Measurement Procedure for Time Delay of
Motor-Overload Relays

Use the following sequence for testing motor overload relays:

1. Set controls to the following:
   •POWER ON/OFF Switch in OFF position (instrument displays off).
   •OUTPUT CONTROL knob at minimum "0" position.
   •OUTPUT MODE switch in center OFF position.
2. Connect one end of a high-current lead to one side of thermal element or current coil in overload relay. Connect other end of this lead to the COMMON terminal of test set.
3. Connect one end of second high-current lead to other side of thermal element or current coil in overload relay. Connect other end of this lead to output terminal labeled 240A.
4. Connect test set to suitable single-phase power supply.
5. Turn test set on with POWER ON/OFF switch (instrument displays should light).
6. Use RANGE SWITCH to select ammeter range, so test current will be near full scale, and no less than 10 percent of full scale.
7. Put ammeter DISPLAY MODE switch in MEMORY position.

8. Connect a pair of light leads (timer leads) from normally closed contacts or normally open contacts of overload relay to binding posts of test set labeled CONTACTS.

9. Select appropriate timer STOP MODE.

10. Select desired timer display mode and range.

11. Rotate OUTPUT CONTROL knob clockwise, momentarily press OUTPUT MODE switch in MOM., and release. Observe current reading retained by ammeter.

12. Continue to rotate OUTPUT CONTROL knob clockwise, while jogging (repeatedly moving to MOM. position and releasing) OUTPUT MODE switch, until desired test current is reached. Suggested test current is three times (3x) the rating of thermal relays, or three times (3x) the pickup current of magnetic relays.

    If desired test current is not reached with OUTPUT CONTROL knob at maximum clockwise rotation, return knob to zero, and transfer output lead from terminal labeled 240A to terminal labeled 120A. Proceed with current adjustment as in Steps 11 and 12.

    If the relay utilizes a high impedance thermal element or operating coil, and the desired test current cannot be reached, transfer output lead to the next higher voltage (lower current) terminal and repeat Steps 11 and 12. If test current is still not reached, transfer output lead to terminal labeled 5A, and repeat Steps 11 and 12.

    NOTE: Before starting test, allow time for thermal element to cool, or in the case of magnetic overload relays, for the piston to reset. Incorrect tripping time may otherwise result.

13. Put ammeter DISPLAY MODE in CONT. position.

14. Start test by moving OUTPUT MODE switch to MAINT. position.

    NOTE: Test current may decrease (fall off) during the test, because the resistance or impedance of the test circuit increases as it heats up. Rotate OUTPUT CONTROL knob clockwise to keep test current at desired value.

15. When overload relay trips, timer stops and output is de-energized. Timer indicates total elapsed time of the test in seconds or cycles.

16. Turn test set OFF with POWER ON/OFF switch.

17. Record test results on Test Record Card.

    NOTE: In order to obtain accurate tripping times with some types of magnetic overload relays, particularly those using high viscosity oil, it may be necessary to "preheat" the relay for a few minutes.

Procedure for Instantaneous Element of Motor-Overload Relays

Use the following sequence to test motor-overload relays.

1. Set controls to the following:

   • POWER ON/OFF switch in OFF position.
   • OUTPUT CONTROL knob at minimum "0" position.
   • OUTPUT MODE switch in center OFF position.

2. Connect one end of a high-current lead to one side of instantaneous element in overload relay. Connect other end of this lead to the COMMON terminal of test set.
3. Connect one end of second high-current lead to other side of instantaneous element in overload relay. Connect other end of this lead to output terminal labeled 240A.
4. Connect test set to suitable single-phase power supply.
5. Turn test set on with POWER ON/OFF switch (instrument displays should light).
6. Use RANGE switch to select ammeter range, so test current will be near full scale, and no less than 10 percent of full scale.
7. Put ammeter DISPLAY MODE in MEMORY position.
8. Connect a pair of light leads (timer leads) from normally closed contacts or normally open contacts of overload relay to binding posts of test set labeled CONTACTS.
9. Select appropriate timer STOP MODE.
10. Select desired timer display mode and range.
11. Rotate OUTPUT CONTROL knob clockwise, momentarily press OUTPUT MODE switch to MOM., and release. Observe current reading retained by ammeter.

    If desired test current is not reached with OUTPUT CONTROL knob at maximum clockwise rotation, return knob to zero, and transfer output lead from terminal labeled 240A to terminal labeled 120A. Proceed with current adjustment as in Step 11.

    If the relay utilizes a high impedance instantaneous element, and the desired test current cannot be reached, transfer output lead to the next higher voltage, lower current terminal, and repeat Step 11. If test current is still not reached, transfer output lead to terminal labeled 5A, and repeat Step 11.

12. Continue Step 11 until overload relay trips. Observe current reading retained on ammeter. Timer indicates elapsed time of test in cycles or seconds.

NOTE: To avoid tripping error caused by interference of time delay element, allow thermal element to cool; or in the case of magnetic overload relays, for the position to reset.

13. Repeat test, starting with OUTPUT CONTROL knob at position just below trip current of instantaneous element observed in Step 12.
14. When overload relay trips, timer stops, and output is de-energized. Current reading is retained on ammeter 1. Timer indicates elapsed time in seconds or cycles.
15. Turn test set OFF with POWER ON/OFF switch.
16. Record test on Test Record Card.

NOTE: Refer to manufacturer's instructions for instantaneous trip time. If increasing test current does not decrease tripping time, current at which minimum tripping time was first observed is the instantaneous trip current value.

MAXIMUM CURRENT OUTPUT

The test set described above will push 750 amperes through the impedance of a 125-ampere breaker. The unit pictured in Figure 9–8 will push 1,500 amperes through a 125-ampere breaker. Units are available that can push 60,000 amperes continuously and 100,000 amperes for instantaneous trip test of power breakers.

SUMMARY

This chapter discusses circuit breaker test sets, and the following major points are made:
The circuit breaker test set is used in the testing of low-voltage circuit breakers. Three tests must be conducted on the low-voltage circuit breakers: one for long time delay, one for short time delay, and one for the minimum operating point of the instantaneous unit. Each pole on low-voltage circuit breakers must be tested individually.

**FIGURE 9–8
Intermediate-sized
circuit breaker test
set. (Courtesy of
AVO Multi-Amp
Corporation)**

REVIEW QUESTIONS

1. Explain how a thermal overload works in a molded-case circuit breaker.

2. In adjusting the magnetic trip of a molded-case breaker, explain what values you are changing.

3. What multiple of current is recommended for the instantaneous test?

4. What is the theoretical time delay for a breaker when it reacts to a fault five times its current rating?

5. What electrical characteristic of the device under test do most high-current test sets rely on to produce the needed current?

6. For most efficient control of test current, which test set terminal should be used first?

7. What are the three tests performed on low-voltage circuit breakers?

8. What advantage does using SCRs instead of contractors have in initiating output of the test set?

9. Why is it necessary to keep test leads as short as possible when using high-current, low-voltage test sets?

10. How can a test set with 15-ampere input produce an output of hundreds of amperes?

Troubleshooting and Testing

A technician is assigned the job of testing a molded-case circuit breaker. Rated at 125 amperes, the breaker is connected to the test set, and three times the rated current is applied. The test set timer shows an elapsed time of 117 seconds.

Test time is so close to the time curves shown in Figure 9–6 (page 220) for this breaker that the technician decides to repeat the test to check the results. An appropriate time for the breaker to cool is allowed, but all attempts to reinitiate the test set fail. A quick check of connections and test set fuses shows no problems.

QUESTIONS
1. Was the breaker good according to the first test results?
2. What typical mistake did the test technician make in repeating the test to the circuit breaker?

ANSWERS
1. Yes. The original test results were within the 10 percent of value shown for manufacturer's curves for this breaker.
2. Remember that the breaker contacts are in series with the load. Under test conditions most new technicians forget to reset the breaker after each test. An open circuit means no current can flow.

Glossary

alternating current Electric current that reverses direction periodically, usually many times per second. Abbreviated AC.

ammeter An instrument for measuring the magnitude of an electric current.

ampere Flow of electrons past a given point in one second. The basic unit of electrical current.

average voltage A method for measuring voltage by taking samples of a number of instantaneous voltages along a sine wave, adding them, and then dividing by the number of samples taken.

capacitor Two conducting materials in close proximity to each other but separated from each other by an insulating material called the dielectric.

current Electron flow through a circuit

cyclometer register A set of four or five wheels numbered from zero to nine inclusive on their edges and so enclosed and connected by gearing that the register reading appears as a series of adjacent digits.

diode A device that allows electrons to flow through it in one direction but blocks the flow in the opposite direction.

direct current An electric current that flows in one direction only. Abbreviated DC.

galvanometer An instrument used to measure small DC currents.

generator A machine that converts mechanical energy into electrical energy.

hertz A unit of frequency equal to one cycle per second.

magnetism Phenomena involving magnetic fields and their effect upon materials.

megohmmeter An instrument that is used for measuring the high resistance of electrical materials of the order of 20,000 megohms at 1,000 volts; one direct reading type employs a permanent magnet and a moving coil.

motor A machine that converts electrical energy into mechanical energy by utilizing forces produced by magnetic field on current-carrying conductors.

null detector A galvanometer or other device that indicates when voltage or current is zero; used chiefly to determine when a bridge circuit is in balance.

ohmmeter A direct reading instrument for measuring electrical resistance.

Ohm's Law The current in an electric circuit is inversely proportional to the resistance of the circuit and is directly proportional to the electromotive force in the circuit.

oscilloscope An instrument primarily for making visible the instantaneous value of one or more rapidly varying electrical quantities as a function of time or of another electrical or mechanical quantity.

parallax The change in the apparent relative orientations of objects when viewed from different positions.

phase angle meter An instrument for measuring the phase displacement, read in degrees, between voltages, currents, or a voltage and a current.

polyphase A circuit consisting of three energized conductors, usually separated from each other by 120 electrical degrees.

potentiometer A three-terminal rheostat, or a resistor with one or more adjustable sliding contacts that functions as an adjustable voltage divider.

power The rate of doing work or the rate of expending energy. The unit of electrical power is the watt.

power factor Efficiency of a circuit or device. It is the ratio of the actual power of an alternating or pulsating current, as measured by a wattmeter, divided by apparent power, as indicated by ammeter and voltmeter readings.

resistance A property of component composition in resisting the flow of current in DC. In AC resistance can be the result of component composition and circuit characteristics.

RMS Root mean square. A method for measuring the efficiency of AC voltage when compared to DC voltage. If both voltages are the same, AC voltage is only 70 percent as efficient as DC voltage. RMS voltage is the figure used in most electrical work.

rotor The rotating member of an electrical machine or device.

selenium rectifiers A metallic rectifier in which a thin layer of selenium is deposited on one side of an aluminum plate and a conductive metal coating is deposited on the selenium.

shunt A device having appreciable resistance or impedance connected in parallel across other devices or apparatus, and diverting some (but not all) of the current from it.

sine wave A wave whose amplitude varies as the sine of linear function of time.

single-phase Energized by a single alternating voltage.

three-phase A combination of circuit energized by alternating electromotive forces, which differ in phase by one-third of a cycle.

transformer A device that, when used, will raise or lower the voltage of alternating current of the original source.

voltage Potential difference or electromotive force measured in volts.

voltmeter An instrument designed to measure a difference in electrical potential, in volts.

volts The unit of electrical potential.

watts Actual work performed by an electrical circuit. The unit of electrical power.

watthour Work performed by an electrical circuit over one hour. The usual billing unit used by the electrical utility.

waveform A manifestation or representation (that is, graph, plot, oscilloscope, presentation, equation, table of coordinate, or statistical data, etc.) or a visualization of wave, pulse, or transition.

Index

Note: Page numbers in **bold type** reference non-text material.